ADVANT

The Future of Immersive Cybersecurity Education

PRACTICAL CYBERSECURITY SERIES

Disaster Response and Recovery

Course Textbook
and Study Guide

MICHAEL I. KAPLAN

Author

SCOTT C. SCHEIDT

Forward

READER FEEDBACK

Phase2 Advantage strives to publish cybersecurity textbooks of the highest standards and quality. Each text is subjected to a thorough development, editing, and review process prior to publication. This process continues throughout the life cycle of the publication.

Reader feedback is a valuable component of this process after the textbook has been published. To submit comments regarding the improvement of the quality of this textbook, contact us via email at *editor@phase2advantage.com*. Please ensure to include the textbook title and ISBN in your email.

TEXTBOOK DESCRIPTION

Business Continuity and Disaster Response and Recovery is the development of processes, policies, and procedures that prepare for and react to significant and unplanned operational disruptions. The *Disaster Response and Recovery* course textbook prepares students to successfully manage Business Continuity Planning and Disaster Recovery missions. Facing daily risks to long-term success from a wide range of threats — cyber-attacks, human error, technical failures, and natural disasters — businesses must create practical plans to sustain their vital operations, security posture, industry reputation, and brand.

Using 14 detailed chapters designed to align with academic calendars, students will cover critical topics such as BCP Design, Risk Management Frameworks, Qualitative and Quantitative Analysis, Asset Inventories and Resource Profiles, Recovery Site Workflows, Cloud Computing Agreements, Cloud Security, and Legal Requirements of Cloud Storage Solutions. Immersive learning labs include Business Impact Analysis case studies and Risk Assessment exercises.

COURSE LEARNING OBJECTIVES

01: Recognize the key components of business continuity and disaster response planning, map planning strategies to organizational objectives, describe appropriate authority documents, list challenges to the desired state of security, and describe the evolution of sustainable policies.

02: Explain the role and importance of a business impact analysis, tangible and intangible costs, data collection methods, BIA program management, key personnel considerations, exception and assumption workflows, and presenting BIA data results to organizational leadership.

03: Perform quantitative and qualitative analysis labs to calculate single and annual loss expectancy, estimate severity and likelihood probabilities, examine asset value considerations, and use sources of respected statistics to create SLE, ALE, and ARO models for the organization.

04: Compare disaster recovery options, describe recovery site management and workflows, and discuss the roles and importance of work area considerations, key personnel selection, validating successful recoveries, and digital communication systems and methodologies.

05: Organize a risk management program strategy focusing on key components such as risk management frameworks, asset inventories and resource profiles, analysis methodologies, vulnerability assessment, cost estimate challenges, and third-party service providers.

06: Evaluate cloud computing service models, architecture and security considerations, risks and threats posed to cloud services, regulatory and compliance requirements, cloud provider and customer responsibilities, and the structure of contracts and service level agreements.

INSTRUCTOR RESOURCES

Training institutions that adopt the *Disaster Response and Recovery* textbook for use in their course curricula may request corresponding instructor resources at no additional cost. These resources include lecture presentation slides, question text banks for each of the 14 chapters, and lab resource guides. For more information please contact Phase2 Advantage.

ADA ACCESSIBLE COURSE MATERIALS

All Phase2 Advantage digital course materials – including textbooks, lab guides, and lecture slides in PDF and PPT formats – are ADA accessible and score **100%** on major Learning Management Systems such as Moodle, Blackboard, Canvas, and LearnUpon. For more information, please visit the Phase2 Advantage website at phase2advantage/higher-education.

ABOUT THE DIRECTOR

Michael I. Kaplan is the Director of Operations for Phase2 Advantage, a cybersecurity training and publishing company based in Savannah, Georgia. He is also the Chairman of the Savannah Technical College Cybersecurity Advisory Committee and heavily involved in curriculum design initiatives. Michael has written numerous courses and cybersecurity training programs for corporate, academic, and government personnel. He has also developed training programs for Law Enforcement and Fugitive Task Force Investigators on the topics of Criminal Topology, Forensic Document Analysis, and Investigations.

Michael's technical areas of specialization are Incident Handling and Response, Network Forensics, Digital Forensics, and Information Technology Risk Management. He also provides consulting services for government, corporate, and academic organizations both domestically and internationally.

You are invited to connect with Michael on LinkedIn:

linkedin.com/in/michaelikaplan

Organizations face ongoing threats to their information technology infrastructure on a daily basis. The *Cybersecurity Defense and Operations* course textbook brings core competencies to advanced levels with new concepts and traditional best practices. Students will cover topics such as Cloud Security, Threat Intelligence, Vulnerability Management, Biometric Systems, Incident Response, Securing Systems with Cryptography, and the NICE Cybersecurity Workforce Framework.

Paperback: 342 Pages
Publisher: Phase2 Advantage
ISBN-13: 978-1737352914

As organizations continue to rely on expanding infrastructure in an increasingly hostile threat landscape, the escalation of incidents poses critical risks to information systems and networks. The *Incident Investigations and Response* course textbook provides students with the ability to identify threats, respond to incidents, restore systems, and enhance security postures. Topics include the Response Life Cycle, Attack Vectors, Investigative Methods, Malware Triage, and Remediation Strategies.

Paperback: 268 Pages
Publisher: Phase2 Advantage
ISBN-13: 978-1737352921

As digital crime increases exponentially, the need for investigative expertise in both government and civilian sectors has increased proportionally as well. The *Network Defense and Investigations* course textbook provides students with methods and strategies to conduct digital investigations in a manner consistent with professional industry standards. Topics include Host and Network Investigations, Memory Forensics, Evidence Collection, Forensic Analysis, Chain of Custody, and Reporting.

Paperback: 324 Pages
Publisher: Phase2 Advantage
ISBN-13: 978-1737352938

PUBLICATIONS AVAILABLE ON VITALSOURCE™

In addition to being available through traditional booksellers such as Amazon and Barnes & Noble, Phase2 Advantage course textbooks are also available in digital format on the VitalSource academic platform owned by the Ingram Content Group.

Founded in 1994, VitalSource is the leading education technology solutions provider committed to helping partners create, deliver, and distribute affordable, accessible, and impactful learning experiences worldwide. As a recognized innovator in the digital course materials market, VitalSource is best known for partnering with thousands of publishers and institutions to deliver extraordinary learning experiences to millions of active users globally. Today, VitalSource is committed to powering new, cutting-edge technologies designed to optimize teaching and learning for the 21st century.

Browse the VitalSource Bookshelf®:

bookshelf.vitalsource.com

CONTACT PHASE2 ADVANTAGE

The Future of Immersive Cybersecurity Education

For additional information regarding course textbooks, immersive learning labs, or instructional design and curriculum development services, please contact our main office via the information provided below.

Phase2 Advantage
P.O. Box 14071
Savannah, GA 31416

Phone:
+1.912.335.2217

Email:
michael.kaplan@phase2advantage.com

Table of Contents

Chapter 14

Knowledge Assessment Answer Keys

Appendix

There is No More Relevant Time Than Now

The year 2020 started out like many years before: everyone coming off a holiday high or, perhaps some would say, a holiday crash, but looking ahead with the normal professional angst of determining new requirements and projects for a new year. In March of 2020, as these requirements and projects were beginning to be implemented, the business world was turned upside down, as the world around us all became entangled in the global Covid-19 pandemic that upended business operations, affected supply lines, and infected the workforce.

For those professionals working in the disaster response and recovery fields, many years had been spent learning about, and preparing for, pandemic impacts on business continuity. Only 30% of small organizations had a business continuity plan in place, compared with 54 percent of medium businesses and 73% of large businesses. Business continuity plans from pre-2020 failed to include little if any information on how the organization would adjust their posture to account for a workforce reduction due to widespread illness.

Traditionally, disaster recovery and response in business continuity planning focuses on what will happen if organizational resources such as buildings or equipment are damaged. An assumption that has been the foundation of business continuity planning is people will be able to return to a specified location, such as a building, after a single event like a utility outage or storm. Next to no thought had been put into sick and infectious workers being unable to report to work for an extended period of time. During the U.S. Covid response plan, businesses, social organizations, and schools were required to close by order of the CDC to help slow the spread of the illness. The implementation of remote workforce policies became critical for many organizations that found themselves with a significantly reduced number of people able to report to work.

I spent 27 years in military service spanning time in the Army Active Duty, Army Reserves, and Army National Guard. I served in garrison environments that supported daily operations of stateside bases, and combat tours in Iraq that spanned three named wartime contingency periods. Response and recovery planning, and continuity of operations, were a regular part of the lives of the soldiers I led and my own professional expectations. We had pandemic response integrated through a highly professional group of Army medical personnel. Commercial and many non-governmental business organizations do not have the luxury of this integrated planning.

While pandemic response and continuity of operations have been the focus of everyone in 2020 and will continue into the near future, it is not the only important critical event that organizations must plan for. Each day news media outlets, both domestically and abroad, are reporting new issues or concerns found by researchers that relate to cybercrime, computer concerns like website defacement, identity theft, malware attacks such as ransomware, and more. We hear of the diversion of funds from organizational bank accounts resulting from phishing emails, and an overabundance of sensitive operational data and intellectual property stolen during cyber-attacks due to poor systems security and a lack of adherence to awareness training received by employees.

Michael I. Kaplan and his team at Phase2 Advantage have done an awesome job at taking the critical information topics related to disaster recovery and response and integrating them into an academic text, which allows both practitioners and new students to understand essential industry topics. The clear focus of topics such as *Business Impact Analysis* (Chapter 02), *Selecting a Risk Management Framework* (Chapter 03), the need to design and implement an *Emergency Operations Center and Plan* (Chapter 05), accounting for *Epidemic and Pandemic* impacts (Chapter 07) on organizational operations, and *Regulation and Compliance* considerations (Chapter 11) that hinder organizational progress, is a well-rounded and refreshing compilation of crucial and highly relevant content required for today's academic institutions to educate students, and for organizations to use to achieve long-term operational success.

It is not a secret that there is a large cybersecurity workforce gap both in the United States and around the world. The NICE Framework is comprised of seven Categories of cybersecurity functions, 33 Specialty Areas of cybersecurity disciplines, and 52 Work Roles that detail the knowledge, skills, and abilities required to perform tasks in the cybersecurity field. One can clearly see where the Phase2 Advantage team has directly aligned this course content with the NICE Workforce Framework for Cybersecurity, most specifically in the category of Protect and Defend, where we find the Specialty Areas of Incident Response and Vulnerability Assessment and Management.

Phase2 Advantage has given us a great asset to help further cyber workforce development, and to integrate understanding of disaster recovery and response modeling that will guide students from their academic experience to successfully applying this knowledge professionally in the cybersecurity industry.

SCOTT C. SCHEIDT

LTC Scott C. Scheidt, MSC, MBA (US Army, Ret.) is the Chief Security Officer for Seimitsu and an adjunct professor in the Cyber and Related Programs Department at Savannah Technical College, Savannah, Georgia. He is a Certified Cybersecurity professional with experience in cyber offensive and defensive operations, vulnerability assessments, risk mitigation, business management, enterprise security, and the development of cyber policy and strategy.

Scott retired from the Georgia Army National Guard in 2019 after leading the GA Cyber Mission Forces, helping to lead the formation of the GA National Guard Cyber Protection Team, and conducting intelligence and cyber operations for over 16 years. LTC Scheidt is well versed in the Cybersecurity Maturity Model Certification (CMMC) framework and helping organizations institutionalize CMMC activities. He is experienced in managing security services for business and government clients as well as developing cyber training plans and curriculum to bridge the cybersecurity workforce shortage gap.

Introduction to Disaster Response and Recovery

KEY KNOWLEDGE POINTS

The BCP and DRP Convergence
Key BCP and DRP Definitions
BCP Key Components
DRP Key Components
Desired State of Security Challenges
The Evolution of Sustainable Policies

What is "Disaster Recovery"?

While the definition of Disaster Recovery differs from entity to entity, the content of this text will be based on the following definition:

Disaster Recovery:

A set of policies, tools, and procedures to enable the recovery and continuation of mission critical technology infrastructure and systems following a natural or human-induced disaster.

Business Continuity / Disaster Recovery Plan Convergence

Business Continuity Planning (BCP) and *Disaster Recovery Planning* (DRP) are both a combination of three primary disciplines with three distinct missions supporting the highest probability of successful BCP/DRP outcomes. These are the three pillars of the "BCP/DRP Tower." Business Continuity and Disaster Recovery plans must account for compliance, risk management, and operational process alignment to be successful.

1. Certification and Accreditation Compliance
2. Risk Management and Audit
3. Business Process Alignment

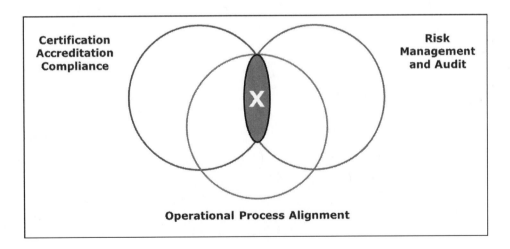

Business Continuity Planning: Key Definitions

There are several key definitions associated with disciplines of BCP/DRP that will be used throughout this text. For the sake of clarity and to reduce confusion, these key definitions have been listed below.

Event:

Any user- or system-generated action or occurrence that can be identified by a program and has significance for system hardware or software.

Incident:

Any unlawful, unauthorized, or unacceptable action that involves a computer system, cell phone, tablet, and any other electronic device with an operating system, or that operates in a computer network.

Maximum Allowable Downtime (MAD):

The absolute maximum time that systems can be unavailable without serious and/or negative impact to the organization.

Recovery Time Objective (RTO):

The targeted duration of time within which a mission-critical business process must be restored after a disaster and/or disruption in order to avoid unacceptable consequences associated with a break in business continuity.

Recovery Point Objective (RPO):

The maximum acceptable amount of data loss, measured in time, that can be incurred without serious and/or negative impact to the organization.

Quantitative Analysis:

The process of collecting and evaluating measurable and verifiable data in order to understand the condition and performance of a business.

Qualitative Analysis:

Examination and evaluation of non-measurable data using subjective judgement and non-quantifiable methods.

Mapping the BCP to Organizational Objectives

Information Security is **not** a business objective in and of itself, but Information Security underlies **all** business objectives. It is important that BCP Managers map their security needs to organizational objectives to get support from senior leadership. A few examples of these objectives are listed below.

1. Maintain Corporate Profit Margin
2. Legal and Regulatory Compliance
3. Maintain a Competitive Advantage
4. Increase and Protect Brand Value
5. Continuous Delivery of Products and Services

Business Continuity Planning (BCP)

The BCP is not a monolithic or linear document. Supported from the top-down and created from the bottom-up, it is a collection of plans that are always evolving. The BCP consists of five primary plans, although the number can vary between entities based on need and context. Each of these plans and their functions will be discussed in further detail later in this chapter.

1. The Administrative BCP
2. The Technical BCP
3. The Work Area BCP
4. The Epidemic / Pandemic BCP
5. The Crisis Management Plan

Disaster Recovery Planning (DRP)

Like its BCP counterpart, the DRP is not a monolithic or linear document and has many moving parts that are executed simultaneously. Additionally, like its BCP counterpart, the DRP consists of five primary plans, although the number can vary between entities based on need and context. Each of these plans and their functions will be discussed in further detail later in this chapter.

1. The Data Recovery Plan
2. The Incident Response Plan
3. The Vulnerability Assessment Plan
4. The Network Forensics Plan
5. The Digital Forensics Plan

Being equally vital, each plan will have a significant part to play in the BCP/DRP mission, and they are not presented here in order of importance. Each will have a significant part to play in the BCP/DRP mission.

The Administrative Continuity Plan

This plan is a consolidated, high-level document written to meet a set of specific business needs, and in a manner that makes sense to the senior leadership. Each plan will change from entity to entity based on organizational needs and context. Listed below are a few common topics that are typically addressed in these high-level plans.

1. **How Company BCP will be Conducted**: This section will outline the overall BCP leadership, strategy, development, and implementation. It will define the scope of the BCP and serve as the written support for the plan by senior leadership.

2. **Long-Term Contingency Planning**: This section will outline long-term contingency strategies based on the vision for the organizational mission.

3. **Common Reference Information**: This section will contain high-level reference information such as organizational charts, list of key vendors, list of key customers, critical assets, and critical personnel as determined in the Business Impact Analysis (to be discussed in a later chapter).

4. **Description of Testing Expectations**: This section will outline the scope and frequency of testing both the BCP and the DRP, the reporting of test results, and the metrics associated with stated goals and successful outcomes.

5. **Keeping Pace with Process Changes**: This section will outline the change management system, version control system, and internal processes that ensure the BCP and DRP remain both relevant and current.

The Technical Continuity Plan

This plan is a detailed document, managed by IT leadership, and written to meet a set of specific technical needs, and in a manner that makes sense to the senior leadership. Each plan will change from entity to entity based on organizational needs and context. Listed below are a few common considerations that are typically addressed in these plans.

1. **Not Restricted to IT**: This plan is compiled and managed by IT, but it is not restricted to IT staff. It must be accessible to anyone expected to be a part of the technical recovery effort.

2. **Addresses All Technical Processes**: Very few leaders are aware of all the behind-the-scenes technical processes that support the vital business functions they manage.

3. **Only Addresses Vital Business Functions**: Only technical processes that support vital business functions should be addressed.

4. **Written by Interdisciplinary Technicians**: This plan is typically written by an interdisciplinary team representing all IT functions in an organization.

5. **Make Process Easy for Technicians**: There is no guarantee the primary technician of a process will be the one tasked with executing elements of this plan. It should be written in such a manner as to allow any competent technician to follow the plan.

The Work Area Continuity Plan

This plan is a detailed document, maintained and managed by BCP leadership, and written to meet a set of specific workflow needs. It focuses on establishing a temporary work area for staff during a recovery process. An IT recovery is useless without an environment that allows key personnel to function in their roles. If staff cannot function, the competition is operating while the organization is not.

It also provides the secondary benefit of enhancing the reputation and brand image of the organization, as customers tend to have more faith in an organization that has planned to continue operating under the most adverse circumstances.

The Epidemic and Pandemic Continuity Plan

This plan is a detailed document written to meet a set of specific staffing needs during the life cycle of an epidemic or pandemic. It is different from other plans in that it addresses an issue that can be foreseen but has a life cycle longer than other disasters. This plan considers several issues applicable to pandemics and serious epidemics.

1. It Affects People at all Levels
2. It Affects Organizations at all Levels
3. It Affects Upstream and Downstream Dependencies

The Crisis Management Plan

This plan is a detailed document written to meet a set of specific response and mitigation needs during an adverse situation. Each plan will change from entity to entity based on organizational needs and context. Listed below are a few common considerations that are typically addressed in these plans.

1. Incident Identification and Reporting
2. Incident Escalation Procedures
3. Incident Handling Policies
4. Stakeholder Notification Procedures
5. Disaster Recovery Initiation Procedures

The Incident Response Plan

This plan is a detailed document written to identify and characterize serious events possessing the capability of negatively impacting business operations. It outlines how leads of value will be defined, how incident timelines will be created, and how investigative priorities will be determined.

This plan guides incident handling personnel in discovering the scope of the incident, and documents policies and standards for the creation and maintenance of case notes. Each plan will change from entity to entity based on organizational needs and context. Listed below are the six common stages of incident response that are typically addressed in these plans.

1. Detection
2. Analysis
3. Containment
4. Eradication
5. Recovery
6. Lessons Learned

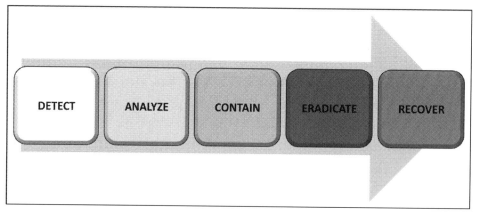

Overview of the Incident Response Strategy

The Vulnerability Assessment Plan

This plan is a detailed document, usually managed by Information Security Management, that is written to discover hidden flaws and weaknesses within critical business systems. Each plan will change from entity to entity based on organizational needs and context. Listed below are five common areas of interest that are typically addressed in these plans.

1. Modeling Attack Chains and Life Cycles
2. Targeted Active Vulnerability Scanning
3. Securely Deploying Architectural Assets
4. Passive Remediation and Containment
5. Offensive Threat Intelligence

The Network Forensics Plan

This plan is a detailed document written to identify the scope and impact of serious network security incidents. A few examples of such incidents are security breaches, data exfiltration, and incidents caused by malicious insiders. Each plan will change from entity to entity based on organizational needs and context. Listed below are five common areas of interest that are typically addressed in these plans.

1. Establishing Live Response Policies
2. Selection of Live Response Tools and Practices
3. System and Data Duplication Priorities
4. Creating Investigative and Analysis Methodologies
5. The Collection and Preservation of Evidence

The Digital Forensics Plan

This plan is a detailed document written to meet a set of specific remediation needs during an adverse situation. A few examples of such situations include malware triage, violations of an acceptable use policy, and utilizing the organizational network for criminal activity. Each plan will change from entity to entity based on organizational needs and context. Listed below are six common areas of consideration that are typically addressed in these plans.

1. Defining Forensic Goals and Objectives
2. Use of In-House or Out-Sourced Expertise
3. Legal Considerations for Network Devices
4. Evidence Collection and Chain-of-Custody Policies
5. Malware Triage and Analysis
6. Notification of External Law Enforcement Agencies

Challenges to the Desired State of Security

The desired state of security is defined as the difference between the current state of security and the state of security that would best protect the goals of the organization. Ideally, the key stakeholders within the organization would share common goals and objective perspectives on the desired state of security. In reality, the desired state must be defined in business and security terms while facing many common obstacles, such as the challenges listed below.

1. **Overconfidence and Unrealistic Optimism**: This is by far the most common issue any BCP Manager must overcome. In the first instance, leadership believes the organization can handle any adverse circumstance with little or no planning. In the second instance, leadership thinks the organization will never experience the adverse circumstance in question. Both share equally bad outcomes.

2. **Psychological Anchoring**: This is a bias present in the decision-making process where one piece of information is given disproportionate consideration and importance. This bias is maintained despite any new information that may challenge the assumption and causes a fixation that excludes consideration of other facts.

3. **Status Quo Bias**: This is another common issue for BCP Managers to overcome and presents itself in phrases such as, "*It has always been done this way; if it is not broken, do not try to fix it.*"

4. **Mental Accounting**: Mental accounting refers to the various values people place on money, based on personal and subjective criteria, that often has detrimental results. An example would be someone who values a "personal fund" of cash in a piggy bank more valuable than cash spent on paying down credit card debt, even though reducing the debt ultimately yields a better financial position than hiding cash.

5. **Herd Instinct**: This behavior is demonstrated when individuals act collectively as a group, with no type of centralized leadership, often in a contrary fashion to their individual beliefs or knowledge. This is also referred to as a "mob mentality."

6. **False Consensus**: This behavior is demonstrated when an individual believes their thoughts to reflect those of others, even if no consensus exists. This is seen primarily in decision-makers.

Defining Authority Documents

Organizational authority, accountability, responsibility, and purpose are defined by authority documents. Although they are formal and definitive, BCP Managers must recognize that organizational culture can create "shadow policies" that serve to diminish or even negate established formal, written authority documents.

There are five general types of authority documents that guide organizational activity, but one must always be aware of the role culture plays in power dynamics.

1. **Policies**: A policy is a deliberate system of principles to guide decisions and achieve rational outcomes. It is typically crafted as a high-level statement of intent and is generally adopted by a governance body within an organization. They can assist in both subjective and objective decision-making processes.

2. **Standards**: A standard is an established norm or requirement for a repeatable technical task. It is usually a formal document that establishes uniform methods, processes, and practices. It may be developed jointly or unilaterally by corporations, regulatory bodies, trade organizations, or government entities. Standards organizations often have more diverse input and usually develop voluntary standards which may later become mandatory if formally adopted by a governing body.

3. **Guidelines**: A guideline is a verbal or written statement by which to determine a course of action. Its general purpose is to streamline processes according to a set of established or best practices. They may be issued by and used by any organization (governmental or private) to make the actions of employees, vendors, or clients more predictable, and presumably of higher quality and consistency.

4. **Procedures**: Procedures are an established or official way of documenting a series of actions conducted in a certain order or manner. They are often instructional, precise, factual, concise, and may be used in both training and the execution of specific functions.

5. **Frameworks**: A framework is a system of rules, ideas, or beliefs that is used to plan or decide something in an effective and logical way. They may be developed for a specific area of interest, such as a dedicated technology implementation, or address a broad area of interest such as improving an organizations' overall security posture. Many governments and industries rely on frameworks for handling regulatory matters, developing flexible and standardized networks of rules and best practices.

The Evolution of Sustainable Policies

Organizational policies are created via an iterative process that is constantly evolving. As BCP policies are created they will experience the same evolutionary process. The first draft of any policy is rarely adequate, and the maturity level of policies tend to be incremental by nature. Below are five common phases policies experience as they mature.

1. **Initial**: Ad Hoc and/or no Formality
2. **Developing**: Informal and Basic Structure
3. **Defined**: Organization-Wide but Lack Controls
4. **Managed**: Defined Roles and Measured Results
5. **Optimized**: Culturally Entrenched and Practiced

Plan Activation Challenges

Challenges to BCP/DRP activation will be discussed in detail in the following chapters, but it is important for BCP Managers to design solutions that account for real-world challenges, not unrealistic solutions created to appease senior management.

In preparation of a detailed treatment of this topic throughout the text, several common plan activation challenges are presented below for consideration.

1. Initial Information Regarding the Disaster is Typically Minimal
2. The Actual Level of Impact is Typically Unknown
3. Access to the Affected Sites can be Restricted or Inaccessible
4. Decision-Making Communications can be Delayed or Disrupted
5. Initial News Reports can be Speculative or Completely False

Introduction to Disaster Response and Recovery

KNOWLEDGE ASSESSMENT QUESTIONS

The following knowledge assessment questions are presented in true / false, multiple choice, and fill-in-the-blank formats. The correct answers are provided in an Answer Key at the end of Chapter 14. These questions may or may not be presented on quizzes and/or tests given by the instructor of this course.

Knowledge Assessment Questions

1) "*A set of policies, tools, and procedures to enable the recovery and continuation of mission-critical technology infrastructure and systems following a natural or human-induced disaster*" is the definition of _____.

A. Business Continuity

B. Incident Response

C. Risk Mitigation

D. Disaster Recovery

2) Which choice below is a goal of information security professionals when mapping the Business Continuity Plan to organization objectives?

A. Increase Employee Benefits

B. Maintain a Corporate Profit Margin

C. Decrease Competitive Advantage

D. Avoid Regulatory Compliance

3) Business Continuity and Disaster Recovery plans must account for compliance, _____, and operational process alignment to be successful.

A. Risk Management

B. Vulnerability Assessment

C. Financial Services

D. Organizational Leadership

4) What statement listed below represents an accurate characteristic of an Epidemic / Pandemic Continuity Plan?

A. Affects People at Some Levels

B. Only Considers Internal Dependencies

C. Different from Other Recovery Plans

D. Affects the Organization at Some Levels

5) "*The maximum acceptable amount of data loss, measured in time, that can be incurred without serious and/or negative impact to the organization*" is the definition of _____.

A. Recovery Time Objective

B. Recovery Point Objective

C. Maximum Allowable Downtime

D. Quantitative Analysis

Knowledge Assessment Questions

6) What statement listed below represents an accurate characteristic of an Incident Response Plan from the perspective of Business Continuity?
A. Details Remediation and Eradication Steps
B. Addresses the Identification of Events
C. Establishes Financial Priorities and Actions
D. Defines the Secure Development Life Cycle

7) The _____ is a consolidated, high-level document written to meet a set of specific business needs.
A. Administrative Continuity Plan
B. Technical Continuity Plan
C. Crisis Management Plan
D. Incident Response Plan

8) What choice below can be described as a documented component of a Vulnerability Assessment Plan within a Business Continuity strategy?
A. Incident Escalation Procedures
B. Malware Triage and Analysis
C. Collection and Preservation of Evidence
D. Threat Intelligence Collection and Analysis

9) The _____ is a detailed document, maintained and managed by BCP leadership, written to address a set of critical workflow needs.
A. Administrative Continuity Plan
B. Digital Forensics Plan
C. Work Area Continuity Plan
D. Technical Continuity Plan

10) What choice below describes a real-world challenge faced by organizations when activating Disaster Response and Recovery Plans?
A. The Level of Impact is Known
B. Initial Information is Typically Minimal
C. Access to Sites can be Unrestricted
D. Initial News Reports can be Informative

True / False Questions

11) Information Security is not a business objective in and of itself, but Information Security underlies all business objectives in every organization.

1. True
2. False

12) The *Business Continuity Plan* (BCP) is a monolithic and linear document. It is supported bottom-up, created top-down, and changes occasionally.

1. True
2. False

13) The *Disaster Recovery Plan* (DRP) is not a monolithic or linear document. It has many moving parts that are executed simultaneously.

1. True
2. False

14) The first draft of a policy is usually adequate, and the acceptance level of policies tends to be widespread by nature and evolve with financing over time.

1. True
2. False

15) There are many common obstacles when defining a desired security state which must be acknowledged and defined in both business and security terms.

1. True
2. False

16) The *Digital Forensics Plan* is a detailed document written to meet a set of specific response and mitigation needs during an adverse situation.

1. True
2. False

17) The *Technical Continuity Plan* is a detailed document, maintained and managed by IT leadership, written to address specific technology needs.

1. True
2. False

True / False Questions

18) Design solutions that account for real-world challenges, not unrealistic solutions which only serve to appease senior management or look good on paper.
1. True
2. False

19) An incident is any user- or system-generated action or occurrence that can be identified by a program and has significance for system hardware or software.
1. True
2. False

20) Qualitative analysis is the examination and evaluation of non-measurable data using subjective judgement and non-quantifiable methods.
1. True
2. False

Conducting a Business Impact Analysis

KEY KNOWLEDGE POINTS

Roles of a Business Impact Analysis
Benefits of a Business Impact Analysis
Managing a BIA Project
Selecting a BIA Project Manager
BIA Data Collection
Presenting Results to Management

The Roles of a Business Impact Analysis (BIA)

1. **Role 01**: The BIA is an analysis of the important functions that are essential to the operation of the business.

2. **Role 02**: The BIA is used to quantify and qualify the value of each function to the business.

3. **Role 03**: The BIA is used to identify the risks posed to the most valuable business functions.

4. **Role 04**: The BIA is used to suggest mitigation actions to reduce the likelihood or impact of identified risks.

5. **Role 05**: The BIA is used to indicate how much is lost per hour or per day for the length of the outage.

6. **Role 06**: The BIA is used to identify the IT systems associated with critical business functions.

The Benefits of a Business Impact Analysis (BIA)

A Business Impact Analysis provides many benefits to the organization, many of which are valuable beyond the scope of a business continuity project. It may identify operational efficiencies or inefficiencies, assets not yet recorded, and provide better insight into the opportunities available for security to align with business processes. Benefits will change from entity to entity based on organizational needs and context. Listed below are seven direct benefits typically provided by a well-conducted BIA.

1. Identify Critical Functions to Protect
2. Identify Tangible Costs of Functions
3. Identify Intangible Costs of Functions
4. Identify Critical Resources for Functions
5. Determine Recovery Time Objectives
6. Identify Vital Business Records
7. Prioritize the Use of Scarce Resources

BIA: Tangible Financial Costs

There are numerous ways the loss of a critical function can have a negative financial impact on the organization. Tangible costs are those which can be accurately measured, calculated, and recorded on a balance sheet. Tangible losses will vary from entity to entity, but they all have one common denominator:

they are real financial losses and objective by nature. Listed below are five examples of tangible losses that may be experienced by an organization.

1. Products Cannot be Shipped
2. Services Cannot be Delivered
3. Increased Waste from Spoilage
4. Penalties Imposed by Customers (SLA Breaches)
5. Legal Penalties for Non-Compliance (Regulatory Issues)

BIA: Intangible Financial Costs

Intangible costs due to the loss of a vital business function can be more difficult to identify but are no less damaging. They may present themselves as delayed costs after the recovery has been successfully executed and vital business functions have been restored. Intangible costs will change from entity to entity based on organizational structures and context. Listed below are five examples of intangible costs an organization may experience.

1. Loss of Customer Goodwill
2. Reduced Confidence in the Marketplace
3. Higher Employee Turnover
4. Damage to the Image and/or Brand
5. Loss of Faith in Senior Management by Stakeholders

Managing a BIA Project

Managing a BIA project is not a simple task that can be assigned to any volunteer. Additionally, the project must be supported both financially and politically from the highest levels of leadership in the organization. Listed below are five high-level functions the BIA project must accomplish to be successful.

1. **Conducted as an Individual Project**: The BIA must be approached as a standalone project and not be added as an afterthought to other existing initiatives.

2. **Approve the Project Budget**: Those responsible for the project must possess the ability to secure financial backing from senior leadership to ensure the highest probability of a successful outcome.

3. **Convey the Importance of Participation**: The project is interdisciplinary by nature and will require participation of key managers and personnel with little time to spare and marginal interest in the project itself. The projects' importance must be conveyed in a manner that convinces everyone they have some type of benefit by participating.

4. **Address Objections and Questions**: Those responsible for the project will be required to answer questions about the projects' scope and purpose and overcome objections from those who are reluctant to reveal what they consider to be sensitive information.

5. **Approve the Report for Submission**: Once the project has been concluded the report must be reviewed, compiled, edited, and approved for submission to senior leadership.

Selecting a BIA Project Manager

A well-run BIA project will build credibility for the overall disaster recovery planning project; a poorly run BIA project will be a disaster unto itself. Selection of the proper project manager is one of the most important decisions in the entire BIA process. The person must possess a high level of integrity as every facet of the organization will be exposed in this process. A significant level of knowledge regarding the organization is needed to understand and know the true value of internal functions and processes. The person must be comfortable moderating discussions between key organizational units, as conversations about a functions' criticality tend to become heated quickly.

BIA Data Collection

An effective data collection process will help quantify the value of each function in terms of financial and legal impacts. This phase in the data collection process is formal, structured, and critical to the success of the BIA project. The collection process typically involves the use of questionnaires, interviews, and training seminars. Collection methods will change from entity to entity based on organizational needs and context. Listed below are seven high-level steps to consider when formulating and initiating the data collection process.

1. **Identify Those Receiving the Questionnaires**: Using an organizational chart, identify the unit managers, subject matter experts, and key personnel who are qualified to provide required information.

2. **Develop Targeted Questionnaires**: Diverse units within the organization will require different questionnaires based on varying functions. The questionnaires must be tailored to specific critical functions and processes.

3. **Provide Training to Respondents**: Conduct training for those individuals who will be receiving questionnaires to explain the intent, purpose, and expectations of the data collection process. Do not assume respondents understand the document that has been created.

4. **Ensure Timely Completion**: Inform the respondents of the date the questionnaires must be completed and do so with the backing of senior leadership. Open-ended projects tend to stay that way for long periods of time.

5. **Review Responses for Clarity**: Just as it is a mistake to assume respondents will understand documents without training, do not assume those reviewing the submitted documents will understand all of the answers. If those reviewing the questionnaires need clarification for a specific response, seek out the respondent and have it clarified.

6. **Conduct Response Review Meetings**: Response review meetings are a good opportunity to get clarification from the respondents and to ask if the respondents need the same for the questionnaires. This holds especially true if a specific set of questions is generating a significant amount of confusion for the respondents.

7. **Compile and Summarize the BIA Data**: Once the data has been collected from the respondents it must be compiled into a manageable format and summarized for senior leadership. This also provides an opportunity to identify disparities in responses provided by different individuals within the same organizational units.

There are several proven strategies to ensure the data collection process proceeds as smoothly as possible. For example, create a targeted questionnaire for one specific organizational unit and distribute it to the chosen respondents. Once they have been developed and tested in a single unit, distribute them to all appropriate business units.

This allows for the questionnaire to be edited and clarified prior to mass distribution. The methods by which questionnaires are developed will change from entity to entity based on organizational needs and context. Listed below are six high-level strategies to consider when streamlining the data collection process.

1. Explain How the BIA Helps all Departments
2. Include a Copy of Executive Support in Writing
3. Request Senior Leadership "Encourage" Respondents
4. Provide Printed Instructions with Questionnaires
5. Provide Examples of Answer Formats
6. Set a Hard Deadline for Completion

Assumption and Exception Processes

There will be occasions in which senior leadership will ask the BIA project manager to make exceptions or assumptions. To prevent the possibility these requests will reflect poorly on the project manager later, it is advisable to establish a procedure by which those exceptions and assumptions are received and incorporated into the project. This formal procedure will document decisions and validate the acceptance of risk by senior leadership. Exception and assumption procedures will change from entity to entity based on organizational needs and context. Listed below are five general steps to consider when creating this formal process.

1. Submit the Request for a Functional Review
2. Conduct an Assumption Security Review
3. Ensure the Assumption Aligns with the BCP Goals
4. Secure Leadership Approval for the Assumption
5. Document all Leadership Approvals in Writing

Identify Appropriate Respondents

The first step in identifying who should receive the questionnaire is to secure a current copy of the organizational chart. The next step is to identify the critical business units and departments within the organization. Once that has been accomplished, the process of identifying appropriate respondents for the BIA questionnaire can begin.

The critical units will change from entity to entity based on organizational needs and context. Listed below are five general guidelines to consider when interacting with potential respondents.

1. **Identify the Leaders of Business Units**: Using a chart of the organizations' unit managers, select the managers to be involved. They will be responsible for ensuring the data collection process is executed within established deadlines.

2. **Leaders are Responsible for Questionnaires**: It is not the function of the project manager to ensure unit staff are completing the questionnaires as required.

3. **Departments Identify Vital Functions**: No one knows the vital function of a specific unit better than those who manage daily. Although the project manager can provide support and guidance, it will be managers and key personnel of those units that identify critical functions.

4. **Resource Requirements for Functions**: Once the unit managers have identified their critical functions, they must then identify the critical resources that allow those functions to operate. These resources and dependencies may be internal or external contingent upon the function.

5. **Include Critical Suppliers if Possible**: Any suppliers that provide a critical service or product for the function must be identified as well. The suppliers do not need to be unique to the unit, but they must be identified

and documented. This could also include technical support for specialized equipment providing a critical service that requires specialized support to sustain its operation.

Often Missed Questions: Personnel

It is important that the project manager not overlook human costs when creating targeted questionnaires. Managers will certainly know the answers to these considerations even though they may be difficult and uncomfortable to address.

1. Employee Morale
2. Employee Turnover (Resignation and Firing)
3. Cost to Hire New Personnel
4. Cost to Train New Personnel
5. Personnel Influence on Social Media

Often Missed Questions: Delayed Costs

Delayed costs are real, tangible, and must be considered when creating the targeted questionnaire. The crisis does not end for an organization immediately after a successful recovery.

1. Wages Paid for no Work
2. Overtime Wages Paid for Recovery Efforts
3. Additional Cost of Financial Credit
4. Devaluation of the Organizations' Stock (if Public)
5. Additional Cost of Insurance

Often Missed Questions: Goodwill

The project manager must ask the right people the right questions and remain aware that biases (discussed previously) may have an impact on responses. It must be stressed to respondents that there are many intangible costs with tangible impacts on their unit.

1. Loss of Shareholder Confidence
2. Loss of Supplier Confidence
3. Loss of Customer Confidence
4. Damage to Organizational Brand and/or Image
5. Loss of Competitive Advantage

Reporting BIA Project Results

Once unit managers have ensured all questionnaires have been returned within the established deadline, the data will be compiled and organized into a hierarchy of targeted reports. The targeted reports will change from entity to entity based on organizational needs and structure. Listed below are five levels of hierarchy to consider when creating targeted reports.

1. Function
2. Work Group
3. Department
4. Organizational Unit
5. Overall Organization

Key Personnel Considerations

When an organization experiences a disaster there will be impacts on key personnel to varying degrees. The impacts can be internal (psychological) or external (geophysical). The project manager may consider the creation of an "Employee Skills Matrix" for the unit leadership.

In the event a key employee expected to execute in the BCP is unable to do so, there may be other employees with unknown skill sets available to fill the gap. The key employee considerations will change from entity to entity based on organizational needs and staffing. Listed below are five issues to consider when identifying key personnel.

1. **Stress Created by the Traumatic Event**: If a person has never experienced a traumatic event there is no way to predict their reaction to it. It is quite possible a traumatic event may render someone psychologically inoperable.

2. **Payment, Housing, and Insurance**: The BCP may consider the payment, housing, and insurance needs for its staff, but have the families of the staff been considered? If the organization secures hotel rooms for displaced staff and their families that do not accept pets – and that family has numerous pets they value highly – what choices will they make, and how will that impact the BCP?

3. **Labor and Talent Management Issues**: Does the BCP anticipate the availability of suitable replacements if key personnel leave the organization as a result of the disaster, or as a result of the organizations' response to a disaster?

4. **Directed to Return but Refuses**: If key personnel are ordered to return by the organization to participate in the recovery process, and they refuse to do so due to family issues, what repercussions will these personnel face? If they are key personnel that cannot be terminated because of a

critical job function, will non-critical personnel be treated in a similar fashion? If non-critical personnel are not treated equally, how will that impact staffing and morale?

5. **National Guard, Firefighters, and EMS:** Personnel may be members of the National Guard, EMS, or volunteer firefighters without revealing that fact to the organization. These issues can be identified in an "Employee Skills Matrix" and address competing loyalties before disaster strikes.

Presenting BIA Results to Management

Once the BIA project has been completed, the corresponding reports must be presented to senior leadership. To ensure the highest probability of a successful outcome, preparation and simplicity are the best methods to present compiled results. Listed below are six presentation tips to consider when making a formal presentation to management.

1. Stay Cool and Start Strong
2. Have a Clear Structure and Objective
3. Know the Stakeholder Audience
4. Anticipate the Tough Questions
5. Make the Message Memorable
6. Ask for What You Need

Slide Presentation Tips

Ideas are judged by how they are presented, and "Death by PowerPoint" should be considered cruel and unusual punishment. It is important for the presenter to remember that **they** are the presentation, not the slides. Be bright, be brief, be memorable, and be gone. Listed below are twelve general tips to consider when using slides in support of formal presentations or training.

1. Limit Bullet Points and Text (5-5-5 Rule)
2. Use High-Quality Graphics
3. Use Appropriate Charts and/or Graphs
4. Use Color Well (Conservative Colors)
5. Use Mainstream Fonts
6. Use a Visual Theme but Avoid Templates
7. Use Scripts and Storytelling
8. Use Dark Text on Light Backgrounds
9. Think Outside the Screen (Slide Fixation)
10. Ask Questions and Engage the Audience
11. Use Voice Modulation and Inflection (No Monotone)
12. Practice the Presentation to Perfection

Conducting a Business Impact Analysis

KNOWLEDGE ASSESSMENT QUESTIONS

The following knowledge assessment questions are presented in true / false, multiple choice, and fill-in-the-blank formats. The correct answers are provided in an Answer Key at the end of Chapter 14. These questions may or may not be presented on quizzes and/or tests given by the instructor of this course.

Knowledge Assessment Questions

1) A _____ provides many benefits to the organization which are valuable beyond the scope of a Business Continuity Planning project.

A. Vulnerability Assessment

B. Financial Projection Analysis

C. Risk Assessment

D. Business Impact Analysis

2) What choice below represents a best practice to be followed when conducting formal presentations at all levels of the organization?

A. Limit Bullet Points and Text

B. Use Creative Fonts

C. Use Popular Templates

D. Use Animated Graphics

3) An effective _____ process will help quantify the value of each function in terms of financial and legal impacts.

A. Qualitative Analysis

B. Asset Distribution

C. Data Collection

D. Quantitative Analysis

4) What choice below represents a potential consequence of a negative impact to an organizations' "goodwill"?

A. Maintaining Competitive Advantage

B. Damage to Brand or Image

C. Increase in Product Revenue

D. Restricted Hiring Budgets

5) Once the Business Impact Analysis _____ have been developed and tested in a single unit, distribute them to all appropriate business units.

A. Reports

B. Questionnaires

C. Budgets

D. Approvals

Knowledge Assessment Questions

6) What choice below represents a best practice to be followed when initially developing a Business Impact Analysis questionnaire?

A. Test Questionnaires in Multiple Units

B. Standardize Questions for Business Units

C. Solicit Feedback and Clarifications

D. Limit Options to Answer Questions

7) Check _____ and travel schedules for all selected respondents to help ensure timely completion of the Business Impact Analysis questionnaires.

A. Vacation

B. Training

C. Meeting

D. Presentation

8) What choice below can be described as a benefit of conducting a Business Impact Analysis in support of a Business Continuity Plan?

A. Identify Unprofitable Functions to Cancel

B. Confirm Recovery Time Objectives

C. Identify and Prioritize Abundant Resources

D. Identify Vital Business Records

9) Create a formal process for exemptions and _____ that documents decisions and validates the acceptance of risk by senior leadership.

A. Key Employees

B. Risk Ratings

C. Questionnaires

D. Exceptions

10) What choice below represents a potential tangible cost to an organization which must be considered when conducting a Business Impact Analysis?

A. Legal Penalties for Non-Compliance

B. Loss of Customer Goodwill

C. Reduced Shareholder Confidence

D. Damage to the Brand Image

True / False Questions

11) To ensure the highest probability of a successful outcome, preparation and simplicity are the best methods to present compiled results to decision-makers.

1. True
2. False

12) Consider creating a temporary staffing list to anticipate and prepare for key personnel considerations which may emerge at the last minute unexpectedly.

1. True
2. False

13) The first step in identifying who should receive the BIA questionnaire is to secure a current copy of the organizational chart.

1. True
2. False

14) Create an informal process for exemptions and exceptions that documents decisions and validates the rejection of risk by senior leadership.

1. True
2. False

15) Once the BIA questionnaires have been developed and tested in a single unit, distribute them to all appropriate business units.

1. True
2. False

16) An effective BIA data collection process will help qualify the value of each function in terms of staffing and productivity impacts.

1. True
2. False

17) The Business Impact Analysis project must be supported financially and politically from the highest levels of the organization.

1. True
2. False

True / False Questions

18) Tangible costs due to the loss of a vital business function can be more difficult to identify but are less damaging than intangible costs.
1. True
2. False

19) There are numerous ways the loss of a critical function can have a negative financial impact on the organization.
1. True
2. False

20) Ensure the right people are asked the right questions regarding goodwill. There are many intangible costs with tangible consequences.
1. True
2. False

Selection of Risk Management Frameworks

KEY KNOWLEDGE POINTS

The Key Attributes of Risk
Risk Management Program Development
Asset Inventories and Resource Profiles
Risk Management Frameworks
Vulnerability Assessment ≠ Risk Assessment
Third-Party Service Providers

Risk Management: Key Definitions

Risk:

A probability or threat of damage, injury, liability, loss, or any other negative occurrence that is caused by external or internal vulnerabilities, and that may be mitigated through preemptive action.

Vulnerability:

The degree to which people, property, resources, systems, and cultural, economic, environmental, and social activity is susceptible to harm, degradation, or destruction on being exposed to a hostile agent or factor.

Likelihood:

The probability of events or situations taking place that have a reasonable probability of occurring but are not definite or may be influenced by factors not yet observed or measured.

Risk Sensitivity:

A relative measurement of a resources' tolerance for risk exposures, independent of any threat or vulnerability.

Risk Analysis:

The process of identifying and analyzing potential issues that could negatively impact key business initiatives or critical projects in order to help organizations avoid or mitigate those risks.

Risk Assessment:

The evaluation and estimation of the levels of risks involved in a situation, their comparison against benchmarks or standards, and determination of the probable severity of their impact.

Risk Appetite:

The level of risk that an organization is prepared to accept in pursuit of its objectives, and before action is deemed necessary to reduce the risk.

Residual Risk:

The amount of risk or danger associated with an action or event remaining after natural or inherent risks have been mitigated by risk controls.

The Threat Landscape

The threat landscape faced by Information Security professions is vast, varied, and growing exponentially in response to advances in technology. Additionally, the complexity of IoT is increasing the attack surface and poses a wide variety of unique risks. Threats will change from entity to entity based on technology assets and functionalities. Listed below are five categories of threats typically considered by BCP managers.

1. Physical and Natural Events
2. Loss of Essential Services
3. Compromise of Information
4. Unauthorized Actions
5. Compromise of Functions

The Key Attributes of Risk

The key attributes of risk are those elements BCP managers must take into consideration when crafting a risk management program. All attributes influence risk – and each other – simultaneously.

Attributes of risk will change from entity to entity based on technology assets and functionalities. Listed below are eight attributes typically considered by BCP managers when developing risk management programs and processes.

1. Warning
2. Scope
3. Predictability
4. Time of Day
5. Time of Week
6. Impact
7. Likelihood

8. Capability

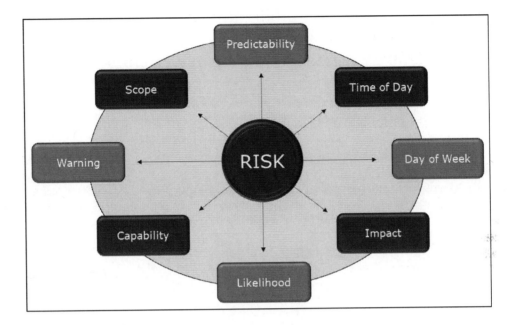

The 5 Layers of Risk

Think of risk in terms of concentric circles, with external risk being the farthest from the center. Remember the influences of the attributes of risk; mentally layering risk helps to prioritize solutions. Listed below are five layers of risk typically considered by BCP managers when developing risk management programs and processes.

1. External Risk
2. Facility-Wide Risk
3. Data Systems Risk
4. Unit and Department Risk
5. Local Proximity Risk (Immediately Observable)

Risk Management Program Development

Developing a risk management program is not complicated, but it is complex insofar as there are many moving parts to consider. As each component is considered, the BCP manager must remain aware of the focus and details of implementation. This focus will help to ensure program development stays within the boundaries of realistic expectations. Listed below are six high-level components typically considered by BCP managers when designing a risk management program.

1. Context and Purpose of the Program
2. Scope, Charter, and Methodology
3. Authority, Structure, and Reporting
4. Asset Identification and Classification
5. Risk Management Objectives
6. Creation of the Implementation Team

The Risk Management Process

The ability to effectively manage risk is contingent upon a successful risk assessment and analysis process. This effort does not have to be created from scratch; there are numerous frameworks available to the BCP manager to support this mission. Listed below are five high-level components typically considered by BCP managers when drafting a risk management process for their organization.

1. Understand the Potential Threats
2. Determine the Existing Vulnerabilities
3. Determine if Risk Levels are Acceptable
4. Assess the Risk Mitigation Options
5. Review the Effectiveness of Implemented Controls

The Risk Management Procedure

The risk management procedure is an ongoing and continual process conducted for every critical asset and process identified within the organization. As the organization evolves and changes, so too will the processes in place to manage new risks. Listed below are five high-level steps in the risk management procedure that tend to stay constant within a changing environment.

1. Identify Information Assets and Their Value
2. Perform a Risk Assessment
3. Determine the Risk Treatment and/or Response
4. Accept or Reject the Residual Risk
5. Monitor the Risk and Communicate Results

Conducting an Asset Inventory

To protect any asset within an organization, the BCP manager must know the asset exists. An asset inventory is a comprehensive list of critical assets that an organization must protect in order to maintain its vital functions and processes. This is important information for the technical BCP that is typically kept in both the BCP and the asset location. The structure of the asset inventory will change from entity to entity based on the type of assets and criticality. Listed below are six categories of information typically considered when creating an inventory.

1. System Type and Version
2. Installed Software (Including Version)
3. Name and/or Title of the Resource Owner
4. Physical and Logical Location
5. Logical Network Addressing
6. Vendor Support Information

Creating an Asset Resource Profile

An asset resource profile is created for each asset listed on the asset inventory and identifies where and how it fits within the risk management strategy. It is supporting information used for risk mitigation efforts that is attached to the inventory documentation. Listed below are six categories of information typically considered when creating a resource profile.

1. General System Description
2. Functions and Features
3. Information / Data Classification
4. Criticality to the Organization
5. Applicable Compliance Regulations
6. Identified User Community

Asset Valuation

Determining the value of an asset can be challenging, as many tangible and intangible factors must be taken into consideration.

Regardless of the challenging nature of this task, this valuation should be included with the asset inventory and resource profile documentation. The valuation methodology will change from entity to entity based on organizational structure and needs. Listed below are five high-level criteria typically considered when conducting an asset valuation.

1. Criticality and/or Sensitivity
2. Cost to Replace
3. Loss of Revenue
4. Legal and/or Regulatory Sanctions
5. Brand Image and/or Reputational Value

Risk Analysis Methodologies

There are numerous frameworks available to the BCP manager, both technical and non-technical, that can serve as a guideline for the development of risk management procedures and programs. There is no "one size fits all" perfect solution; these models should be adapted and utilized based on specific organizational culture and needs. Listed below are four methodologies that have gained widespread appeal and acceptance: OCTAVE Allegro, FRAAP, FAIR, and the NIST RMF. This list is not meant to be all-inclusive.

1. **OCTAVE Allegro**: The *Operationally Critical Threat, Asset, and Vulnerability Evaluation* uses three phases, and a highly detailed questionnaire to assist in identifying risk. First, an asset-based threat profile is determined. Next, vulnerabilities and exploitation methodologies are identified. Finally, risks and potential mitigation strategies are identified using this information. This methodology is best suited for small projects and one-time assessments.

2. **FRAAP**: The *Facilitated Risk Analysis and Assessment Process* uses subject matter experts and a streamlined approach to conduct risk analysis for project-level assessments. Interdisciplinary experts, led by a facilitator, meet for 4 to 8 hours to conduct the analysis. This method encourages collaboration and can generate a final report in a few days. FRAAP is also easily adaptable into other risk assessment methodologies.

3. **FAIR**: The *Factor Analysis of Information Risk* is based on two factors to measure risk: *Loss Event Frequency* (LEF) and *Probable Loss Magnitude* (PLM). Using measurable statistics of frequency, capability, and control strength it creates a six-level risk exposure table that can be used as the source of mitigation strategies. This is a detailed quantitative and probabilistic analysis method best suited for larger and more advanced projects.

4. **NIST RMF**: The *National Institute of Standards and Technology Risk Management Framework* is a real-time continuous monitoring risk analysis methodology. It supports the use of automated tools such as *SIEM* (Security Information and Event Management) to inform risk decisions, integration with system architecture, and integration at all levels of the System Development Life Cycle. It also establishes a model for the accountability of security controls. The NIST RMF uses a six-step, high-level approach that consists of the following categories:

    a. **Categorize**: System Information
    b. **Select**: Security Controls
    c. **Implement**: Security Controls
    d. **Assess**: Security Controls
    e. **Authorize**: Information System
    f. **Monitor**: Security Controls

Vulnerability Assessment ≠ Risk Assessment

A vulnerability assessment is a useful tool to identify potential weaknesses that may be visible to threat sources, but it is not the same as a risk assessment. Listed below are five facts about vulnerability that distinguishes it from risk.

1. Used to Identify Potential System Weaknesses
2. Based on Rules, Known Exploits, and Parameters
3. Not all Vulnerabilities Pose Risks
4. One Vulnerability can Pose Many Risks
5. A Component of the Risk Assessment Process

Risk Assessment ≠ Vulnerability Assessment

A risk assessment is a useful tool to evaluate risk and develop mitigation strategies that align with organizational processes, but it is not the same as a vulnerability assessment. Listed below are five facts about risk assessment that distinguishes it from vulnerability assessment.

1. Based on Category of Information and/or Data
2. Accounts for System Architecture
3. Accounts for Existing Security Controls
4. Based on Capability, Likelihood, and Impact
5. Provides a Multi-Dimensional View of the Environment

Manufactured Risks: Cost Estimate Challenges

Risk can be avoided, assumed, transferred, and mitigated but it cannot be removed entirely. Nowhere is this more apparent than with manufactured risks. These risks are difficult to quantify and qualify, the BCP manager must be aware that they exist even if they cannot be controlled. Listed below are five examples of these risks that expose many organizations to potential disasters daily.

1. **Industrial Sites**: Toxic Chemical Tanks
2. **Airports**: Organizations Located in Flight Paths
3. **Fires**: Water and Smoke Damage Impacts
4. **Pipelines**: Combustible Gas and/or Fuels
5. **Infrastructure**: Fragility and/or Dependencies

Third-Party Service Providers

Security managers should be involved in all phases of the contracting process and ensure supporting providers have adequate controls. Although this topic will be covered in detail in Chapter 13, "*Cloud Computing Contracts and Service Agreements*," a high-level overview is warranted in this discussion as outsourcing contracts tends to increase potential risk.

Not all third-party service providers can be audited, but for those that can, the audit requirement should be negotiated into the *Service Level Agreement* (SLA). Listed below are several factors to consider when utilizing third-party service providers.

1. Ensure the Contracts Specify the Required Security Levels
2. Ensure a Risk Assessment is Performed on the Third-Party Company
3. Require that Post-Relationship Processes be Followed
4. Ensure all Regulatory Requirements are Managed and Enforced
5. Require the Rights to Source Code Placed in Escrow
6. Require the Vendors' Obligation to Remain Compliant
7. Secure the Right to Audit the Vendor Process (if Possible)
8. Insist on the Use of Standard Operating Procedures
9. Ensure the Right to Assess the Skillsets of any Resources Provided

Selection of
Risk Management Frameworks

KNOWLEDGE ASSESSMENT QUESTIONS

The following knowledge assessment questions are presented in true / false, multiple choice, and fill-in-the-blank formats. The correct answers are provided in an Answer Key at the end of Chapter 14. These questions may or may not be presented on quizzes and/or tests given by the instructor of this course.

Knowledge Assessment Questions

1) "*A probability or threat of damage, injury, liability, loss, or any other negative occurrence that is caused by external or internal vulnerabilities, and that may be mitigated through preemptive action*" is the definition of _____.

A. Risk

B. Vulnerability

C. Likelihood

D. Severity

2) Which of the choices listed below is a component and best practice of the risk management process?

A. Understand Previous Threats

B. Determine Potential Vulnerabilities

C. Accept the Levels of Risk

D. Review Control Effectiveness

3) "*The evaluation and estimation of the levels of risks involved in a situation, their comparison against benchmarks or standards, and the determination of the probable severity of their impact*" is the definition of _____.

A. Risk Appetite

B. Risk Assessment

C. Vulnerability Assessment

D. Residual Risk

4) Which of the choices listed below is a characteristic to consider when creating an asset resource profile?

A. System Type and Version

B. Logical Network Addressing

C. Functions and Features

D. Vendor Support Information

5) Although risk management program development consists of 6 high-level steps, do not lose focus or awareness on the eventual details and procedures of _____.

A. Approval

B. Financing

C. Implementation

D. Acceptance

Knowledge Assessment Questions

6) Which of the choices listed below is a characteristic of the OCTAVE Allegro risk analysis framework and methodology?

A. Requires Pre-Planning by a Facilitator

B. Uses a Three-Phased Approach

C. Based on Two Factors to Measure Risk

D. Establishes Accountability for Controls

7) A(n) _____ provides critical information for the technical BCP and is typically kept in both the BCP and the asset location.

A. Resource Profile

B. Personnel Log

C. Gantt Chart

D. Asset Inventory

8) Which of the choices listed below is a characteristic of the Facilitated Risk Analysis and Assessment Process (FRAAP)?

A. Focused on Project Level Assessments

B. Real-Time Continuous Monitoring

C. Creates a 6-Level Risk Exposure Table

D. Uses Highly Detailed Questionnaires

9) The identification and costs of _____ risks is highly beneficial to a thorough risk analysis. Be aware these risks exist even if they cannot be controlled.

A. Manufactured

B. Transferred

C. Mitigated

D. Residual

10) Which of the choices listed below is a characteristic of the Factor Analysis of Information Risk (FAIR) framework and methodology?

A. Building Asset-Based Threat Profiles

B. Automated Tools to Inform Decisions

C. A Streamlined Approach Using SME's

D. Probable Loss Magnitude (PLM)

True / False Questions

11) Security managers should be involved in all phases of the SLA process and ensure supporting providers have adequate controls.

1. True
2. False

12) Risk management frameworks and analysis methodologies provide value to selected processes if their primary focus is technical and data driven.

1. True
2. False

13) When choosing a risk management framework, adapt and utilize established reference models based on organizational culture and specific needs.

1. True
2. False

14) Risk assessment considers the basic factors of asset value, both tangible and intangible, within several contexts and potential scenarios.

1. True
2. False

15) An asset resource profile is supporting confidential information for risk mitigation. Document the resource profile with the asset inventory information.

1. True
2. False

16) There are numerous frameworks and industry tools available for assessing risk, but the components of vulnerabilities and threats share many similarities.

1. True
2. False

17) Risk appetite is the level of risk that an organization is prepared to accept in pursuit of its objectives, and before action is necessary to reduce the risk.

1. True
2. False

True / False Questions

18) Risk analysis is a relative measurement of a resources' tolerance for risk exposures, independent of any threat or vulnerability.

1. True
2. False

19) Not all third-party service providers can be audited, but for those that can, the audit requirement should be negotiated into the Service Level Agreement (SLA).

1. True
2. False

20) Risk assessment is a useful tool to identify weaknesses that may be visible to threat sources.

1. True
2. False

Qualitative and Quantitative Analysis Measurements

KEY KNOWLEDGE POINTS

Qualitative and Quantitative Analysis
Defining Severity
Estimating Severity and Likelihood
Asset Value Considerations
Calculating SLE, ARO, and ALE
Sources of Respected Statistics

An Overview of Qualitative Analysis

Qualitative Analysis:

Examination and evaluation of non-measurable data using subjective judgement and non-quantifiable methods.

Most qualitative analysis approaches use a relative scale to rate risk exposures based on a set of predefined criteria for each level. Like all analysis methodologies, it has strengths and weaknesses. Its strength lies in affording BCP managers a way to measure risks that cannot be measured quantitatively and map this risk to a final exposure value. However, this method does have limitations and weaknesses. Listed below are three primary shortcomings to consider when using this method for evaluating risks.

1. It Relies Heavily on the Knowledge of the Assessor
2. It is Subjective and Prone to Inaccuracies
3. It Uses a Scale Based on Descriptive Criteria

Defining Severity

The severity rating used in the qualitative analysis method is meant to describe the scope of the exposure, not list all the potential consequences. It is asset agnostic and measures the magnitude of exploiting a weakness contingent upon its impact on the CIA (or CIAA) Triad. The severity is rated by the degree of disruption based on the time and scope of the resources that are affected. Although there can be any number of levels in a severity rating, the rating scales are typically low, medium, and high.

Data Availability Severity

Developing qualitative risk scales for data availability severity are a great opportunity to design a business focus directly into the risk model. After all, an organization that cannot access or disseminate its data cannot execute its mission. The goal for the BCP manager is to create a clear set of criteria for each level of severity with visible distinctions between each level.

If the distinction between the levels proves to be obscure and confusing, the severity scales can be revised for clarity. Data severity scales will change from entity to entity based on organizational needs and context. Listed below are four metrics typically considered by BCP managers when creating these scales.

1. Impact on External Services
2. Impact on Internal Processes
3. Degradation of Overall Performance

4. Length of time of the Disruption

Data Integrity Severity

Data integrity severity will focus primarily on unauthorized or unintended access to create, read, update, or delete data ("CRUD"). Additionally, the severity scales will consider the varying degrees of unauthorized or unintended access. Integrity severity scales will change from entity to entity based on organizational needs and context. Listed below are three examples of how varying degrees of severity can be used by the BCP manager to create these qualitative scales.

1. **Low**: May Indirectly Contribute to Integrity Issues
2. **Medium**: May Allow Limited Unauthorized Access
3. **High**: May Allow Unrestricted Unauthorized Access

Questions for Estimating Severity

Estimating severity using a qualitative analysis method may prove to be challenging for BCP managers. If difficulty exists qualifying the severity of a vulnerability, many useful questions can be asked to improve clarity. Listed below are six examples of questions that can be asked by the BCP manager to create these scales.

1. What is the Scope of Severity Post-Exploitation?
2. What is the Degree of Service Shutdown?
3. Will its Form be Executable Code or Arbitrary Functions?
4. Will Accessed Data Help Exploit Other Systems?
5. Will the Impact be Internal or External?
6. Are These Unique or Aggregated Threats?

Questions for Estimating Likelihood

Estimating likelihood using a qualitative analysis method can pose similar challenges for BCP managers as when estimating severity. As with estimating severity, if difficulty exists qualifying the threat/vulnerability pairs, many useful questions can be asked to improve clarity. The BCP manager must also be aware that it is not uncommon to confuse the contributing factors impacting the likelihood and severity of the threat/vulnerability pairs. Listed below are eleven examples of questions typically asked by the BCP manager to create these scales.

1. What is the Size and/or Population of the Threat Universe?
2. Is There a Location Requirement for Exposure?
3. Is There Available Information About the Exploit?

4. What Skill Level is Required to Execute the Exploit?
5. How Attractive is the Target to Malicious Actors?
6. Has the Exploit Been Executed in the Past?
7. Is the Exploit Applicable in the Current Environment?
8. Is an AV, IDS, or IPS Signature Available?
9. Is Authentication Required for the Exploit?
10. What is the Impact on Servers and/or Endpoints?
11. Is the Vulnerability Widely Deployed?

The Challenges of Quantitative Analysis

Quantitative Analysis:

The process of collecting and evaluating measurable and verifiable data in order to understand the condition and performance of a business.

Quantitative analysis methods are more exact than their qualitative counterparts insofar as they are an objective means to evaluate measurable and verifiable data. Although this can be true in many situations, this method has challenges.

Many quantitative models have been proposed over the years with complex equations for calculating risk, but none have been the "silver bullet" for BCP managers. Listed below are six examples of challenges typically faced when using this method.

1. **Lack of Accessible Historical Data**: This method relies on historical data to create benchmarks and predict trends. If such data does not exist or is not accessible, originating assumptions will be speculative.

2. **Reliance on Extremely Accurate Data**: If historical data is available and accessible it must also be extremely accurate for the analysis to yield effective results.

3. **Requires Significant Time to Analyze**: Large datasets take a significant amount of time to analyze and may not be practical for short-term needs. Even with the support of machine learning systems, human participation to evaluate results cannot be avoided.

4. **Reluctance to Share Vital Information**: Accurate historical data may be available and accessible to members of a specific organization, but few organizations are inclined to share their vital information with external entities. This can also occur within a single organization if the culture is compartmentalized.

5. **Anonymous Surveys are Rarely Honest**: In the absence of accurate historical data, anonymous questionnaires may be used to collect needed information (Delphi Method). In many cases respondents will not provide answers that are truthful for fear their responses will be traced back to them.

6. **Data is Skewed to Specific Industries**: Large studies generating a high volume of statistics may provide useful information but tend to be skewed to the specific industries that sponsored or commissioned the study.

Asset Value Considerations

The valuation of assets, both tangible and intangible, is a critical component of risk assessment, analysis, and management. The assessment must transcend line-item numbers as the asset value can extend well beyond the actual cost.

Asset value considerations will change from entity to entity based on organizational needs and context. Listed below are six questions typically asked by the BCP manager to determine the true value of an organizational asset.

1. What is the Actual Cost of the Asset?
2. What is the Cost to Replace if it is a Legacy Asset?
3. What is the Functional Value to the Organization?
4. Do any Operational Dependencies Exist?
5. What are the Assets' Compliance Requirements?
6. What is the Criticality of the Assets' Data?

Calculating SLE, ARO, and ALE

The BCP manager can use qualitative analysis to determine asset value by calculating Single Loss Expectancy and Annualized Loss Expectancy for each critical asset. Review the definitions below to become familiar with the related terms listed below.

Exposure Factor (EF):

The subjective, potential percentage of loss to a particular asset if a specific threat occurs.

Single Loss Expectancy (SLE):

The monetary value expected from the one-time occurrence of a risk on an asset.

Annual Rate of Occurrence (ARO):

An estimated probability of frequency of an occurrence of a risk on an asset each year.

Annualized Loss Expectancy (ALE):

The monetary value expected from an occurrence of a risk on an asset distributed over a given period of time.

Calculating Annualized Rate of Occurrence (ARO)

The ARO is a method for calculating probability over time and is one of the components used in the ALE equation. To determine ARO divide the number of years by the yearly likelihood of an occurrence as shown in the table below.

1 Year	1/1	**1.0**		6 Years	1/6	**.17**
2 Years	1/2	**0.5**		7 Years	1/7	**.14**
3 Years	1/3	**.33**		8 Years	1/8	**.13**
4 Years	1/4	**.25**		9 Years	1/9	**.11**
5 Years	1/5	**0.2**		10 Years	1/10	**0.1**

Calculating Single Loss Expectancy (SLE)

To calculate SLE, multiply the Asset Value by the Exposure Factor. This operation is expressed as AV x EF = _____. This answers the question, "*What is the assets' value and what percentage of the value will be impacted by an occurrence?*" For example, an asset with a value of $5,000 that has a 75% exposure will have a SLE of $3,750 (5,000 x .75 = $3,750). The EF exists because an occurrence may not have a 100% impact on a given asset.

Calculating Annualized Loss Expectancy (ALE)

To calculate ALE, multiply the Single Loss Expectancy by the Annual Rate of Occurrence. This operation is expressed as SLE x ARO = _____. This answers the question, "*What is the yearly cost of an occurrence on an asset if the cost is distributed over a given number of years?*" Using the numbers from the example above, an asset with a SLE of $3,750 ($5,000 x 75%) would have an ALE of $1,875 ($3.750 x .5) if the asset were exposed to an occurrence, and the losses from that occurrence were distributed over a two-year period.

This qualitative method can be used with any asset, and Exposure Factor, and any Annual Rate of Occurrence. However, the BCP manager must be aware that the costs generated by this method may be underreported due to the assets' data and delayed costs.

To effectively calculate the true ALE of a given asset other factors must be taken into consideration. Still using the numbers from the examples above, assume the asset was a server used for email. The ALE was calculated as $1,875. Now assume the asset is a database server in a hospital containing *Personally Identifiable Information* (PII) and *Personal Healthcare Information* (PHI).

Does the $1,875 calculation for ALE remain correct? Listed below are a few questions to consider when answering that question.

1. What is the average cost per file for a data breach?
2. What are the regulatory fines for a data breach?
3. What are the legal costs arising from civil litigation?
4. What is the cost to provide credit reporting for victims?

Asset Value: The Broader Picture

The *Ponemon Institute* (https://www.ponemon.org), a respected source of data statistics, estimates that the average cost per file in 2018 compromised in a breach was $158. The regulatory fines for a data breach vary between industry sectors but can be obtained from the regulatory agencies themselves. The legal costs for civil litigation cannot be quantitatively determined, but a wide variety of respected sources provide accurate historical data and trends.

The typical annual cost for credit monitoring and reporting is $99. Although only two values are definitively known, a new ALE can be calculated that presents a different picture.

Using the ALE value from the examples above, calculate the knowable cost of a $5,000 (AV) database server with a 75% exposure factor (EF) containing 10,000 customer files with PII. The original ALE of $1,875 would still be accurate but the values listed below must now be added.

1. **Files Lost:**   7,500   (10,000 x .75)
2. **File Cost:** $1,185,000   (7,500 x $158)
3. **Reporting:** $742, 500   (7,500 x $99)
4. **Litigation:**   Unknown
5. **Fines:**   Unknown
6. **Known Total:**   **$1,927,500** (not $1,875)

Many organizations take a line-item approach to calculating the value of assets and, for the majority of circumstances, that approach is rational. Unfortunately, it can present an inaccurate assessment for senior leadership and an underfunded program for a BCP manager. Both the senior leadership and the BCP manager will be unprepared and ill-prepared in the event a response to a disaster is required.

Sources of Respected Statistics

The BCP manager has access to respected sources of statistical data that will provide information to "fill in the blanks" and create more accurate assessments for senior leadership. Listed below are a few sources which are discussed in this text, but this is by no means meant to be an all-inclusive list.

The Ponemon Institute

Organizations engage *Ponemon Institute* to conduct studies on topics that support their thought leadership and marketing objectives. The Institute is best known for its annual Cost of Data Breach sponsored by IBM and the annual Encryption Trends study now sponsored by n-Cipher. Other topics, to name a few, include the cost of insider risks, endpoint security, the economics and effectiveness of security operation centers, how to prepare for a data breach, the importance of prevention in the cybersecurity lifecycle, application security, vulnerability management, third parties and the IoT risk and privileged access management.

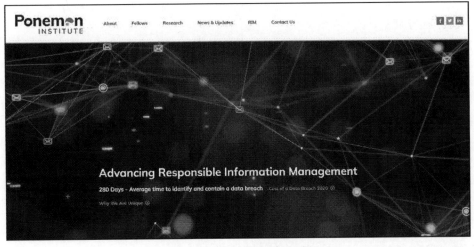

https://www.ponemon.org

Common Vulnerabilities and Exposures (CVE)

The mission of the *CVE Program* is to identify, define, and catalog publicly disclosed cybersecurity vulnerabilities. There is one CVE Record for each vulnerability in the catalog. The vulnerabilities are discovered then assigned and published by organizations from around the world that have partnered with the CVE Program.

https://cve.mitre.org

The OWASP Foundation

The *Open Web Application Security Project*® (OWASP) is a nonprofit foundation that works to improve the security of software. Through community-led open-source software projects, hundreds of local chapters worldwide, tens of thousands of members, and leading educational and training conferences, the OWASP Foundation is the source for developers and technologists to secure the web.

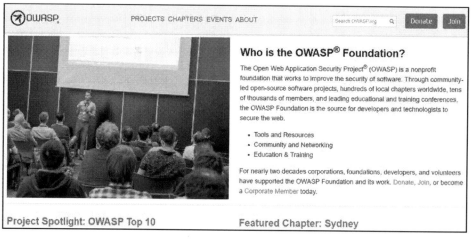

https://www.owasp.org

Qualitative and Quantitative Analysis Measurements

KNOWLEDGE ASSESSMENT QUESTIONS

The following knowledge assessment questions are presented in true / false, multiple choice, and fill-in-the-blank formats. The correct answers are provided in an Answer Key at the end of Chapter 14. These questions may or may not be presented on quizzes and/or tests given by the instructor of this course.

Knowledge Assessment Questions

1) Most _____ analysis approaches use a relative scale to rate risk exposures based on a set of predefined criteria established for each level.

A. Qualitative

B. Predictive

C. Statistical

D. Quantitative

2) Which of the choices listed below is a component used in the calculation of Single Loss Expectancy (SLE)?

A. Exposure Factor

B. External Impact

C. Annualized Rate of Occurrence

D. Annualized Loss Expectancy

3) It is not uncommon to confuse factors that will affect the severity and likelihood of the threat / _____ pair.

A. Response

B. Exploitation

C. Vulnerability

D. Risk

4) *"An analysis method which uses subjective judgment to analyze a company's value or prospects based on non-quantifiable information"* is the definition of _____.

A. Risk Analysis

B. Qualitative Analysis

C. Threat Analysis

D. Quantitative Analysis

5) The severity rating, commonly used in qualitative analysis models, is rated by the degree of potential _____.

A. Risk

B. Disruption

C. Likelihood

D. Consequences

Knowledge Assessment Questions

6) Which of the choices listed below is a factor to be considered when rating the severity of data availability?

A. Deletion of Data

B. Modification of Data

C. Degradation of Performance

D. Unauthorized Access

7) When defining risk scales for data availability severity it is important to ensure there are visible distinctions between _____.

A. Levels

B. Assets

C. Threats

D. Vulnerabilities

8) *"An analysis method of a situation or event, especially a financial market, by means of complex mathematical and statistical modeling"* is the definition of _____.

A. Risk Analysis

B. Qualitative Analysis

C. Threat Analysis

D. Quantitative Analysis

9) The OWASP Foundation is an online community that produces freely available methodologies, tools, and technologies in the field of _____ security.

A. Physical

B. Database

C. Corporate

D. Web Application

10) Which of the choices listed below is a factor to be considered when rating the severity of data integrity?

A. Modification of Data

B. Length of the Disruption

C. Single Loss Expectancy

D. Annualized Loss Expectancy

True / False Questions

11) An accurate assessment must transcend intangible risk severity factors as the value of the asset may extend well beyond the actual financial cost.

1. True
2. False

12) Single Loss Expectancy (SLE) is multiplied by the Annualized Rate of Occurrence (ARO) to determine Annualized Loss Expectancy (ALE).

1. True
2. False

13) Most quantitative analysis approaches use a relative scale to rate risk exposures based on a set of predefined criteria established for each level.

1. True
2. False

14) The severity rating is meant to describe the scope of the exposure, not list all the potential consequences resulting from the exposure.

1. True
2. False

15) Developing qualitative risk scales that address data availability are a great opportunity to design a threat assessment focus directly into the risk model.

1. True
2. False

16) Integrity severity concerns will focus primarily on unauthorized or unintended access to create, read, update, or delete data ("CRUD").

1. True
2. False

17) It is not uncommon to confuse factors that will affect the severity and likelihood of the risk exposure value.

1. True
2. False

True / False Questions

18) An assets' Exposure Factor (EF) is multiplied by the Asset Value (AV) to determine Single Loss Expectancy (SLE).

1. True
2. False

19) The mission of the Common Vulnerabilities and Exposures (CVE) Program is to conduct empirical studies on critical IT infrastructure issues.

1. True
2. False

20) The primary purpose of data integrity severity ratings is to address varying degrees of unauthorized access to data.

1. True
2. False

CHAPTER 05

Creating an Emergency Operations Plan

KEY KNOWLEDGE POINTS

The EOC in Limited Emergencies
EOC Scope and Purpose
EOC Material Requirements
EOC Command Functions
EOC Control Functions
Mobile EOC Characteristics

Disaster Scenario: Chaos

Imagine an organization without a comprehensive BCP/DRP attempting to deal with the aftereffects of a major disaster. Large groups of people are working randomly with no leadership on projects unrelated to actual needs. Without communication, there is no focus on mission-critical priorities and critical actions are overlooked. Repetitious tasks are being initiated and unknowingly undone, and no one at the scene is aware of the project status.

During this chaos, the leadership is demanding updates every hour. That is not a good scenario for any organization to experience. Unless leaders can communicate and direct their staff during emergency situations, they are unable to effectively lead, and this scenario might very well become the reality.

Disaster Scenario: Control

Now imagine an organization with a comprehensive BCP/DRP and established procedures for initiating an *Emergency Operations Center* (EOC) attempting to deal with the aftereffects of a major disaster. Everyone knows where to report when the disaster strikes. Team leaders document the availability of personnel and assign tasks based on their areas of expertise. All recovery actions are directed by a single person prepared to manage the task.

Team locations and their composition are noted, relief teams are seamlessly integrated, and the project status board is updated regularly. The EOC allows organizational management to reestablish leadership, allocate resources, and focus on containment and recovery. That would be a better scenario for an organization to face.

The Emergency Operations Center (EOC) in Limited Emergencies

Disasters can vary in scope and impact. However, an EOC still serves the purpose of a rally point during limited emergencies and short-term, contained disasters. The EOC is typically a place that "everybody knows, and everybody goes," and makes sense for the given situation. It can be a security office with a radio network, or a help desk center with access to the network.

No matter what venue is chosen, the EOC should have a contact number everyone would think to call during an emergency. Additionally, the EOC must be pre-established, pre-supplied, with a location known by everyone before it is needed.

EOC Scope and Purpose

EOC scope and purpose will change from entity to entity based on organizational structure and needs. When it is activated, there are two primary teams: the

Containment Team and the Recovery Team. Listed below are five primary functions typically considered by BCP managers when planning the creation of an EOC.

1. Reestablish Organizational Leadership
2. Facilitate the Allocation of Resources
3. Focus on Emergency Containment
4. Initiate the Disaster Recovery Process
5. Minimize Disruption of Management

EOC Characteristics

No two EOC's are identical but most share similar characteristics. Most organizations convert an existing facility with specific capabilities into an EOC as needed. EOC characteristics will change from entity to entity based on organizational structure and needs. Listed below are six primary characteristics typically considered by BCP managers when planning the creation of an EOC.

1. Typically Range from 500 to 2,000 Square Feet
2. Easily Accessible by Road
3. Ready Access for Delivery Services
4. Proximity to Food Service (Dine-In and Take-Out)
5. Proximity to Hotels and Lodging
6. Wired for Data and Communication

EOC Material Requirements

A disaster is not the time to determine what you need in an EOC; careful planning in advance will help ensure faster recovery. EOC material requirements will change from entity to entity based on organizational structure and needs.

Listed below are six basic categories of supplies typically considered by BCP managers when planning the material requirements for an EOC.

1. Electricity (Generator and UPS)
2. Emergency Lighting
3. Available Sanitary Facilities
4. Medical Kits and Supplies
5. Pre-Packaged Office Supplies
6. Bottled Water and Non-Perishable Food Items

Uninterruptable Power Supply (UPS) Considerations

Prior to sizing electrical support units such as generators and UPS, the BCP manager must know what they will need to support in the EOC. This sounds like a commonsense statement but in many cases this fact is overlooked. Selection of proper UPS systems will change from entity to entity based on EOC structure and needs. Listed below are five general guidelines typically followed by BCP managers when considering UPS requirements for an EOC.

1. Use a UPS with a Dual-Inverter for Smoother Transition
2. UPS Battery Size: Watts Usage x 1.6 (WPVA)
3. Do not Use a Generator to Charge the UPS Battery
4. Use a UPS with Logs to Analyze Noise, Sags, and Spikes
5. Establish a Power Shedding Priority Plan for the UPS

EOC Communications

As stated earlier in this chapter, unless leaders can communicate and direct their staff during emergency situations, they are unable to effectively lead. The Command Center of the EOC will communicate with teams, news media, vendors, customers, the community, and a broad range of stakeholders. Effective communication is critical to control the organizational message and manage perceptions.

Selection of communication systems will change from entity to entity based on organizational structure and EOC needs. Listed below are six basic categories of communication typically considered by BCP managers when planning an EOC.

1. Telephones (Multiple Lines)
2. Radio Communication (Field Crews)
3. Data Communications
4. Organizational Website (Lightweight Pages)
5. Human Messengers (Sensitive Information)
6. Television and AM/FM Radios

EOC Key Personnel

Given the interdisciplinary mission of the EOC, key personnel that represent critical units and functions will be members of the team. Additionally, an important staffing consideration is for every member of the EOC to have a predesignated and cross-trained backup when possible. There will be times when key personnel expected to report to the EOC will be unable to do so. The more practical reason for cross-trained backup is that key personnel need to rest and refresh themselves.

The composition of key personnel will change from entity to entity based on organizational structure and EOC needs. Listed below are six categories of personnel typically considered for the staffing requirements of an EOC.

1. **Recovery Site Manager**: This individual represents the senior leadership of the organization and has the ultimate authority to direct all activities at the site. These responsibilities include progress reporting, requests for supplies, oversight of site staff assignments and safety, and validating the successful recovery of systems as they occur.

2. **Facility Engineering Manager**: This individual is responsible for the mechanical, electrical, and logistical services of the organization. Areas of oversight and responsibility include HVAC systems, electrical and plumbing systems, maintenance of operational systems, and identifying the resources required to facilitate needed restorations.

3. **Public Relations Manager**: This individual is responsible for controlling and disseminating the messaging of the organization based on the needs and direction of senior leadership. Areas of oversight and responsibility include organizing press conferences, preparing media kits, and countering any negative publicity which may arise as a result of the disaster and subsequent recovery effort.

4. **Information Technology Manager**: This individual is responsible for all the technological, networking, and corresponding support systems of the organization. Areas of oversight and responsibility include validating the successful recovery of critical systems, ensuring the implementation of network security, managing the continued operation of functioning systems, and directing technology staff at the recovery site per the needs and direction of the Recovery Site Manager.

5. **Human Resources Manager**: This individual is responsible for all the employee staffing, compliance, and policy enforcement requirements of the organization. Areas of oversight and responsibility include assisting in the staffing needs of the recovery site, tracking employee payroll, and ensuring compliance regarding the health and welfare of the recovery site personnel.

6. **Security and/or Risk Manager**: This individual is responsible for the management of security programs and the continued mitigation of risk to the organization during the disaster and subsequent recovery efforts. Areas of oversight and responsibility include establishing and directing security procedures at the recovery site, ensuring employee safety, controlling access to the recovery site, and directing security and risk management staff at the recovery site per the needs of the Recovery Site Manager.

EOC Command Functions

The EOC has two primary functions during a crisis: *Command* and *Control*. If the EOC does not take control of the decision-making process, people may tend to make expensive and potentially hazardous decisions. The EOC command functions will change from entity to entity based on organizational structure and situational needs. Listed below are five EOC command functions typically executed during disasters.

1. **Gather Damage Assessments**: It is important to understand the scope of the disaster and the subsequent impact on the organization prior to initiating any recovery efforts. This can also be problematic as initial information can be incomplete or inaccurate. Gathering assessments that are relevant and accurate will save time and effort during the recovery process.

2. **Develop Action Plans**: Based on the assessments that are gathered and analyzed, an action plan must be developed to contain the damage and recover the critical functions of the organization.

3. **Allocate Resources Efficiently**: Scarce resources must be allocated efficiently to ensure critical functions can be restored quickly. This can be difficult when many requests for resources flow into the EOC simultaneously.

4. **Approve Deviations from the BCP**: Even the best battle plans must be adjusted in the fog of war and the BCP is no exception. Key personnel with a high-level view can allow changes to be made to BCP procedures as required and take the responsibility for justifying those decisions.

5. **Prioritize Responses to Circumstances**: When planned responses lead to unplanned circumstances, changes to the pre-planned priorities will inevitably occur. Key personnel in the EOC can alter response priorities and have the authority to ensure the changes are communicated.

EOC Control Functions

The control function involves obtaining and dispatching resources based on the direction of the EOC manager. The EOC control functions will change from entity to entity based on organizational structure and situational needs. Listed below are five EOC control functions typically executed during disasters.

1. Ordering Supplies and Services
2. Track Recovery Efforts and Recovery Personnel
3. Implement the Allocated Resources Efficiently
4. Gather Data and Maintain Status Reporting

5. Control the Flow of Information (Internal and External)

Mobile EOC Characteristics

A *mobile EOC* is a viable option for larger organizations with many geographically dispersed facilities. Instead of creating a recovery site for each location a mobile EOC can be dispatched to the area in which the organization was impacted. This tends to be a much more cost-effective solution than multiple recovery sites. Vehicles used as an EOC tend to be large—such as a recreational vehicle or camping trailer—and are equipped to provide comfort for extended lengths of time.

This vehicle is preloaded with everything necessary to establish an EOC, including a generator and tent for expanding the work area. The mobile EOC configuration will change from entity to entity based on organizational structure and geographical needs. Listed below are five characteristics typically associated with mobile EOC's regardless of their size, make, or model.

1. Pre-Supplied with Generators and Tents
2. Requires Cellular Capability for Voice and Data
3. Digitized Floor Plans with Wiring Drawings
4. Door Keys for Critical Access Points
5. Security Credentials for Mobile Response Staff

Creating an
Emergency Operations Plan

KNOWLEDGE ASSESSMENT QUESTIONS

The following knowledge assessment questions are presented in true / false, multiple choice, and fill-in-the-blank formats. The correct answers are provided in an Answer Key at the end of Chapter 14. These questions may or may not be presented on quizzes and/or tests given by the instructor of this course.

Knowledge Assessment Questions

1) A(n) _____ Plan allows organizational management to reestablish leadership, allocate resources, and focus on containment and recovery.

A. Technical Continuity

B. Incident Response

C. Vulnerability Assessment

D. Emergency Operations

2) Which of the choices listed below can be considered a primary responsibility of an Emergency Operation Center's command function?

A. Gather Damage Assessments

B. Order Supplies and Services

C. Implement Allocated Resources

D. Track Recovery Personnel

3) When an Emergency Operations Center is activated, there are two primary teams: the _____ Team and the Recovery Team.

A. Investigative

B. Remediation

C. Containment

D. Logistics

4) Which of the choices listed below is a general material requirement that should be included when planning an Emergency Operations Center?

A. Funding for Overtime

B. Pre-Packaged Office Supplies

C. Human Resource Policies

D. Employee Immunizations

5) A disaster is not the time to determine what items are needed in an Emergency Operations Center; _____ will help ensure a faster recovery.

A. Hiring Policies

B. Careful Planning

C. Outsourced Solutions

D. Purchased Facilities

Knowledge Assessment Questions

6) Which of the choices listed below is an important factor to be considered when determining uninterruptible power supply (UPS) needs?

A. Logs to Analyze Data Events

B. Generators to Recharge the UPS

C. Battery Size and Watts Usage

D. Power Increase Plans

7) During an emergency, the _____ will communicate with teams, news media, vendors, customers, the community, and a broad range of stakeholders in the organization.

A. Command Center

B. Board of Directors

C. Senior Leadership

D. Local Managers

8) Which of the choices listed below typically falls within the scope and purpose of an Emergency Operations center?

A. Reassign Organizational Leadership

B. Initiate the Incident Investigation

C. Form the Disaster Recovery Plan

D. Minimize the Disruption of Management

9) The EOC control function involves obtaining and allocating _____ based on the needs determined by the EOC manager and leadership staff.

A. Damage Assessments

B. BCP Deviations

C. Action Plans

D. Resources

10) Which of the choices listed below represents a potential characteristic of a poorly managed disaster response scenario?

A. Large Groups Working Randomly

B. Status Board is Updated Regularly

C. Action is Directed by a Single Person

D. Everyone Knows Where to Report

True / False Questions

11) The mobile EOC vehicle is preloaded with everything necessary to establish an EOC, including a generator and tent for expanding the work area.
1. True
2. False

12) The EOC command function involves obtaining and allocating resources based on the needs determined by the EOC manager and leadership staff.
1. True
2. False

13) An important staffing consideration is for every member of the EOC to have a predesignated and cross-trained backup whenever possible.
1. True
2. False

14) Prior to sizing electrical support units such as generators and UPS, planners must know what financial requirements they will need to support in the EOC.
1. True
2. False

15) Most organizations convert an existing facility with specific capabilities into an Emergency Operations Center as needed.
1. True
2. False

16) An Emergency Operations Center must be pre-established, pre-supplied, and be an undisclosed location that is known only by senior EOC personnel.
1. True
2. False

17) The mobile EOC vehicle is preloaded with everything necessary to establish an EOC, including a generator and tent for expanding the work area.
1. True
2. False

True / False Questions

18) If key stakeholders do not take control of the decision-making process, expensive and potentially hazardous decisions may be made in the absence of leadership.

1. True

2. False

19) An Emergency Operations Plan allows organizational management to reestablish leadership, allocate resources, and focus on containment and recovery.

1. True

2. False

20) Unless leaders can secure budgets and hire new staff during emergency situations, they are unable to effectively lead and manage the situation.

1. True

2. False

CHAPTER 06

Recovery Site Management and Workflows

KEY KNOWLEDGE POINTS

Assembly Point Considerations
The Recovery Site Manager
Recovery Gantt Chart Functions
Work Area Considerations
Digital Communication Considerations
SMS Notification Methodologies

The Disaster Recovery Site: Key Definitions

Cold Site:

A *Cold Site* is a business location that is used for backup in the event of a disruptive operational disaster at the normal business site and typically does not have the necessary equipment to resume prompt operations.

Warm Site:

A *Warm Site* is a business location that is used for backup in the event of a disruptive operational disaster at the normal business site and is typically suited for bringing up non-essential systems that do not require the immediate restoration that mission-critical systems do.

Hot Site:

A *Hot Site* is a business location that is used for backup in the event of a disruptive operational disaster at the normal business site and is typically a fully operational commercial disaster recovery service that allows continuity of operations in a short period.

Reciprocal Site:

A *Reciprocal Site* is a business location shared by multiple organizations that is used for backup in the event of a disruptive operational disaster at the normal business site and is typically suited for similarly configured organizations with similar recovery requirements.

Assembly Point Considerations

A conditioned response is critical to effective crisis management; time for recovery is decreased when teams can be assembled and tasked quickly. This capability requires the selection of a predetermined assembly point for all personnel involved in either the containment or recovery mission. Assembly point considerations scenarios will change from entity to entity based on organizational structure and needs. Listed below are seven factors typically considered by BCP managers that can influence potential locations.

1. Location During Work Hours
2. Location Outside of Work Hours
3. Access to Data and Telephone Services
4. Access to Wireless and Internet Services
5. Well Lit at all Hours
6. Easy for Personnel to Locate
7. Ample Parking Away from Main Entrance

The Recovery Site Manager

The *Recovery Site Manager* represents the senior leadership and must be able to manage technical and non-technical activities. The individual must also have the authority to make critical decisions without the need of leadership approval. The requirements and selection criteria will change from entity to entity based on organizational structure and needs. Listed below are five primary responsibilities typically assigned to the Recovery Site Manager.

1. Assigns Tasks to all Personnel
2. Oversees the Tracking of Personnel
3. Maintains the Recovery Activity Log
4. Validates Successful Recoveries
5. Directs Status Reporting (Incoming and Outgoing)

Assigning Personnel Tasks

When a recovery site is activated following a disaster there is no time to argue about job boundaries or authority. Site personnel must do what they are instructed to do, when they are instructed to do it, even when the tasks fall outside their job description. It is not uncommon for IT personnel to push-back when assigned non-technical jobs such as sanitation or food delivery. Although the range of tasks and assignments is broad and far-reaching, there are five primary categories of tasks the Recovery Site Manager must consider.

1. Assign and Ensure Site Security
2. Appoint an Alternate (Backup) Site Manager
3. Assign Status Reporting Duties
4. Assign Shifts and Publish a Rest Plan
5. Reassign Employees as Required

Recovery Site Personnel Tracking

Knowing who is on-site ensures safety and efficiency of recovery site staff and provides documentation acknowledgement after a successful recovery. If a smart card access system is operational at the site, the logs will be generated electronically. If not, a roster kept by security personnel at the site entrance can be used. The means and methods of personnel tracking will change from entity to entity based on organizational structure and needs. Listed below are five typical objectives associated with personnel tracking.

1. Facilitates Employee Assignments
2. Helps Avoid Wasting Time
3. Identify the Need for Rest Periods
4. Account for Missing or Overdue Staff
5. Logs are Useful for Post-Recovery Actions

The Recovery Activity Log

The *recovery site activity log* is used to record significant events during the recovery and helps with post-recovery analysis and planning. It is usually maintained by one individual assigned by the Recovery Site Manager. The policies and procedures associated with this log will change from entity to entity based on organizational structure and needs. Listed below are five typical categories of information tracked by the activity log.

1. Mandatory Task Initiation Reporting
2. Mandatory Task Completion Reporting
3. Requests for Service and Supplies
4. Status Reports to and from the Command Center
5. A Source of Information for Gantt Charts

The Function of a Recovery Gantt Chart

The *recovery site Gantt chart* is publicly posted, easy to read, and answers the question, "*When will it be ready?*" It is usually used for hour-by-hour status reporting and serves to identify progress delays in advance. The time frames and formats of Gantt charts will change from entity to entity based on organizational structure and needs. Listed below are five typical characteristics associated with a Gantt chart and its use.

1. Tracks Hour-by-Hour Recovery Status
2. Creates a Logical Sequence of IT Recovery
3. The Chart is Created During BCP Testing
4. The Chart Compares Estimated and Actual Times

5. Identify Recovery Delays and Predicts Completion Times

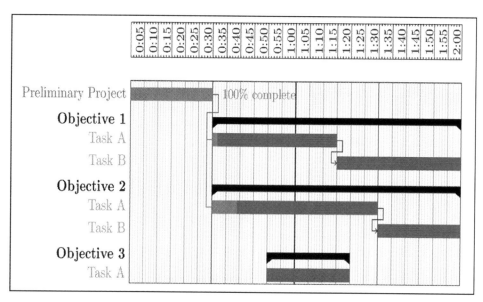

Sample Recovery Site Gantt Chart

Validating Successful Recoveries

Validating successful recoveries provides a layered testing strategy that is useful for catching errors and reducing the time and need to troubleshoot issues repeatedly. Locating the specific technician originally responsible for recovering the system in question—who may already be involved in another recovery effort—can prove to be troublesome. The process by which successful recoveries are validated will change from entity to entity based on organizational structure and needs. Listed below are the five typical phases associated with validating recoveries.

1. The Application and/or System is Restored
2. It is Tested by the Initial Technician
3. The Functionality is Tested by a "Power User"
4. The Recovered System is Released for General Use
5. The Command Center is Notified of the Success

Recovery Site Work Area Considerations

Designing the recovery site work area is a team effort that requires critical departments to be involved in every step of the process. It involves much more than the placement of desks, chairs, and filing cabinets. The BCP manager must also take into consideration such factors as building codes and occupancy limits that remain in effect even during disasters. The process for planning work areas will change from entity to entity based on organizational structure and needs. Listed below are the four typical characteristics to consider when planning a recovery site work area.

1. Plan for 70 to 80 Square Feet per Person
2. The Square Footage Accounts for all Building Space
3. Design Work Areas to be 36" W x 24" D
4. Rotate Shifts to Maximize Work Area Space

Team Member Seating

Team member seating is a critical consideration when planning the recovery site work area that prevents "first come, sit anywhere" chaos. People tend to be conscious of seating arrangements, and even more so when confidential communication is required. The process for planning team member seating will change from entity to entity based on the physical work area and organizational needs. Listed below are five guidelines typically considered by the BCP manager when planning team member seating arrangements.

1. Collocate Interactive Teams (i.e., Finance and Payroll)
2. Recognize that Human Resources and Legal Need Privacy
3. Label Every Desk by Department
4. Label Every Cabinet by Content
5. Suspend Location Signs Conspicuously from the Ceiling

Direct Status Reporting

One individual (and a backup) will be directed to be responsible for maintaining all status reporting between the disaster location and the recovery site. Structured and effective communication is a two-way street that is critical to reduce the chaos and confusion caused by stressful environments. The timeframes and format of status reports will change from entity to entity based on the direction of the Recovery Site Manager and organizational needs. Listed below are five areas of communication typically included in all status reporting initiatives.

1. Status of the Restoration Timeline
2. Names of Personnel at the Recovery Site
3. Supplies and/or Services Required by Recovery Site
4. Progress Updates from the Disaster Location
5. Status of Ongoing Recovery Site Requests

Backup Tape Transport Considerations

Although organizations are trending in the direction of moving backup data from magnetic media to a cloud solution, many legacy systems still rely on tape for data backup. Long-term storage using a professional service does not usually raise concerns, but the transport of magnetic backup media to the recovery site presents specialized and high-risk challenges for the BCP manager.

The tape storage and transportation methods will change from entity to entity based on organizational technology and needs. Listed below are five areas of concern the BCP manager should consider when planning the transport of magnetic media to a recovery site.

1. Magnetic Tape is Degraded by Temperature Change
2. Tape can be Damaged by Sunlight and/or Air Particles
3. Always Handle Magnetic Tape with Lint-Free Gloves
4. Tape can be Damaged by Electronics and/or Machinery
5. Tape can Become Outdated (and Useless) Quickly

Data Deduplication

Deduplication of data provides an efficient compression, single instance storage solution that reduces required storage space and recovery time. Source-based deduplication occurs before backup, allows for less storage space, but may create bottlenecks in slower systems. Target-based deduplication occurs at the backup facility, tends to be much faster, but requires a larger amount of storage space which may be underutilized.

Digital Communication Considerations

When planning for specialized issues associated with the restoration of digital communication, the BCP manager must document alternative communication methods in the event planned methods prove to be ineffective. Communication strategies and priorities will change from entity to entity based on organizational technology and needs. Listed below are five areas of concern typically considered by BCP managers when developing contingency plans for digital communication needs.

1. Note the Location of Wiring Closets and Patch Panels
2. Ensure Route Separation for 100% Redundancy
3. Map Internal and External Cabling Routes
4. Conduct an Inventory of Communication Assets
5. Identify Communication Restoration Priorities

"Call-Tree" Challenges

Despite advances in communications technology, there are a few organizations that still employ the "call-tree" methodology for mass internal notifications. Given current alternatives, this method for team member notification is prone to error and an inefficient option at larger scales. "Call-Trees" are better than nothing, but not by much.

Mass communication strategies and methodologies will change from entity to entity based on organizational policy and needs. Listed below are six areas of concern a BCP manager should take into consideration when contemplating the use of this method for mass notification during disasters.

1. People on the List may be Difficult to Reach
2. The Person may Neglect to Call Others on the List
3. Some People do not Respond to Unknown Numbers
4. Some People Turn Off Work Phones After Hours
5. Cellular Networks can Become Overloaded in Disasters
6. Curiosity and Conversations Slow Down the Processes

SMS Notification Systems

Short Message Service (SMS) notifications are an efficient mass communication method that allow the same message to be broadcast via text to everyone at the same time. This method is much more efficient than the "call-tree" system for mass internal notification.

SMS strategies and implementations will change from entity to entity based on organizational technology and needs. Listed below are five general benefits the BCP manager should consider when planning the use of SMS for notification strategies.

1. Virtually Every Mobile Device can Receive Text
2. Text Messages are Read More Quickly
3. SMS Utilizes Little Amounts of Bandwidth on Networks
4. SMS Works if Data Network is Overloaded
5. SMS does not Require the Internet to Function

SMS Notification Methodology

Although internal recipients of mass notifications can receive a text message at the same time, different levels of management and team members with different roles will not need to hear the same messages.

The verbiage of messages sent to various personnel will change from entity to entity based on organizational policy and a listing of multiple tiered contacts in the SMS system. Listed below are the four tiers of organizational personnel a BCP manager should consider when developing an SMS notification strategy.

1. Executives and Senior Management
2. Functional and Unit Management
3. Line Managers and Supervisors
4. "All-Call" General Notification to Organizational Personnel

Recovery Site Management and Workflows

KNOWLEDGE ASSESSMENT QUESTIONS

The following knowledge assessment questions are presented in true / false, multiple choice, and fill-in-the-blank formats. The correct answers are provided in an Answer Key at the end of Chapter 14. These questions may or may not be presented on quizzes and/or tests given by the instructor of this course.

Knowledge Assessment Questions

1) "*A business location that is used for backup in the event of an operational disaster at the normal business site and typically does not have the necessary equipment to resume prompt operations*" is the definition of a _____ site.

A. Hot

B. Reciprocal

C. Warm

D. Cold

2) Which of the choices listed below is a factor to be considered when planning team member seating needs at disaster recovery sites?

A. Suspend Location Signs from the Ceiling

B. Disperse Interactive Teams

C. Label Every Desk by Employee Name

D. Label Every Cabinet by Function

3) A _____ is critical to effective crisis management; time for recovery is decreased when teams can be assembled and tasked quickly.

A. Training Plan

B. Recovery Site

C. Conditioned Response

D. Awareness Plan

4) Which of the choices listed below is a risk to be considered when planning for the transportation of magnetic backup media?

A. Recovery Site Location

B. Temperature Change

C. Date of Data Storage

D. System to be Restored

5) The _____ Manager represents the senior leadership and must be able to manage technical and non-technical activities effectively under pressure.

A. Risk Assessment

B. Recovery Site

C. Information Technology

D. Physical Security

Knowledge Assessment Questions

6) Which of the choices listed below is a method of data deduplication which is typically conducted at the backup recovery site?

A. Source-Based

B. Inline

C. Target-Based

D. Post-Process

7) "*A business location that is used for backup in the event of a disruptive operational disaster at the normal business site and is typically a fully-operational commercial disaster recovery service that allows continuity of operations in a short period of time*" is the definition of a _____ site.

A. Hot

B. Reciprocal

C. Warm

D. Cold

8) Which of the choices listed below makes SMS notification systems a preferred choice of efficient mass communication?

A. Every Laptop can Receive Texts

B. High Bandwidth Usage on the Network

C. Requires the Internet to Function

D. Text Messages are Read Quickly

9) Validating _____ provides a layered testing strategy that is useful for catching errors and reducing the need to troubleshoot issues in the future.

A. Key Personnel

B. Operational Plans

C. Response Policies

D. Successful Recoveries

10) Which of the choices listed below is a standard function and purpose of an activity log at a disaster recovery site?

A. Requests for Service and Supplies

B. Status Reports to the News Media

C. An Alternative to the Gantt Chart

D. Optional Task Initiation Reporting

True / False Questions

11) Different levels of management and team members with different roles will need to hear the same broadcast messages with the same level of detail.

1. True

2. False

12) The "call tree" method for team member notification is prone to error and an inefficient option at larger scales. For this reason, it is rarely used during a crisis.

1. True

2. False

13) When planning for specialized issues associated with the restoration of digital communication, document alternative and practical sources of funding.

1. True

2. False

14) Deduplication of data provides an efficient compression, single instance storage solution that reduces required storage space and speeds up recovery time.

1. True

2. False

15) The transport of digital backup media – although no longer common - presents specialized and high-risk challenges for the recovery site team.

1. True

2. False

16) Structured communication is a two-way function that is critical to reduce the chaos and confusion which exists naturally in stressful environments.

1. True

2. False

17) The recovery site Gantt chart is considered confidential, easy to understand, and answers the question *"who is managing the project?'*

1. True

2. False

True / False Questions

18) Knowing who is on-site ensures the safety and efficiency of recovery site staff and provides documentation for post-recovery acknowledgement.

1. True
2. False

19) The recovery site activity log is used to record all events during the recovery and helps with pre-recovery analysis, planning, and crafting budget requests.

1. True
2. False

20) The Recovery Gantt Chart is used for daily status reporting and serves to identify and validate successful progress in the recovery effort.

1. True
2. False

Preparing for
Epidemics and Pandemics

KEY KNOWLEDGE POINTS

Seasonal Flu vs. Pandemic Flu
Flu Epidemics: Facts and Statistics
Influenza Questions and Answers
The Epidemic / Pandemic BCP Team
Epidemic / Pandemic Risk Assessments
Post-Epidemic / Pandemic Considerations

Epidemics and Pandemics: Key Definitions

Epidemic:

A widespread occurrence of a disease in a community that spreads quickly and affects many individuals at the same time.

Pandemic:

An epidemic that becomes very widespread and affects a whole region, a continent, or the world due to a susceptible population and causes a high degree of mortality.

Seasonal Flu vs. Pandemic Flu

There are several high-level factors and characteristics which differentiate *seasonal influenza* from *pandemic influenza*. Within each category there will also be distinctions – such as strain, virulence, origin, etc. – but the characteristics listed below are good benchmark indicators of the influenzas' type.

Seasonal Influenza	Pandemic Influenza
Predictable Patterns	Unpredictable Patterns
Occurs Annually	Occurs Rarely
Some Immunity Exists	No Preexisting Immunity
Vaccines Initially Available	No Vaccines Initially Available
Healthy Patients Lower Risk	Healthy Patients High Risk
Modest Impact on Society	Major Impact on Society

6 Phases of a Pandemic

The *World Health Organization* (WHO) has divided pandemics into six phases; the phase will dictate when a disaster is declared. The phase at which point a disaster declaration is made will change from entity to entity based on organizational structure and needs. When the spread of an influenza strain reaches "Phase 5" - Country to Country (Region) – the WHO will usually issue a formal declaration that the world is experiencing a pandemic.

1. Animal to Animal
2. Animal to Human
3. Human to Human (Limited)

4. Human to Human (Community)
5. Country to Country (Region)
6. Country to Country (Global)

Flu Epidemics: Facts and Statistics

Despite the distinction of definitions between epidemics and pandemics, an epidemic is simply a localized pandemic. Impacts of seasonal influenza on an organization can be severe. Listed below are six epidemic statistics for the BCP manager to consider when creating the Epidemic / Pandemic BCP. The statistics have been assembled by the *Centers for Disease Control and Prevention* and represent 2017-2018 averages (https://www.cdc.gov/flu).

1. Flu is a Contagious Respiratory Illness
2. Flu is a Virus-Based Infection (not Bacteria)
3. Flu is Most Contagious in the First 3-4 Days
4. Infection Vectors: Eyes, Nose and Mouth
5. Infection Radius: Up to 6 Feet
6. There are 9.3 – 49 Million Flu Illnesses per Year

Flu Epidemics: Vulnerable Populations

When an Epidemic / Pandemic BCP is written it is important not to forget the family members of organizational staff. In addition to the possibility that they may be a member of a vulnerable population, a family members' illness can keep healthy staff away from their organization. Listed below are six characteristics of vulnerable populations during influenza epidemics compiled by the *Centers for Disease Control and Prevention* (https://www.cdc.gov/flu).

1. Children Younger than Age 5
2. Adults Older than Age 65
3. Pregnant Women (Up to 2-Weeks Postpartum)
4. People with Weakened Immune Systems
5. People with Chronic Illnesses
6. People with a Body Mass Index ≥ 40

Influenza Questions and Answers

According to World Health Organization statistics (*see the link listed below*), seasonal influenza epidemics occur worldwide annually. In temperate climates, they tend to occur in the winter months. The amount of time from incubation to illness is short – only two days – after which point vaccinations become largely ineffective against that strain of virus.

Most healthy individuals recover from an influenza virus within two weeks. However, reported statistics indicate approximately three million to five million severe cases are reported annually, leading to an estimated 290K to 650K respiratory deaths within the same time period.

Given the statistics provided by both the *World Health Organization* (WHO) and the *Centers for Disease Control and Prevention* (CDC) it becomes apparent as to why epidemic and/or pandemic contingencies should be factored into the BCP / DRP planning documentation.

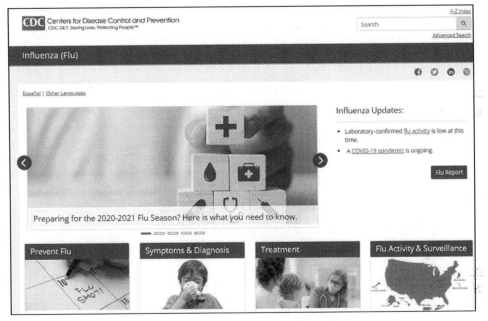

https://www.cdc.gov/flu

Employee Flu Vaccinations

Providing flu vaccinations to organizational staff has proven to significantly lower the impacts experienced by the organization and the spread of the disease between staff. However, no one can be compelled to submit to a mandatory vaccination. The availability of immunization plans for staff will change from entity to entity based on organizational policy and employee need.

The five statistics below representing 2017-2018 averages have been assembled by the organization *Healthline* (https://www.healthline.com) and should be considered by those incorporating a vaccination plan into their Epidemic / Pandemic BCP (https://www.vaccines.gov/get-vaccinated/where).

1. Delivered by Injection or Nasal Spray
2. Average Cost to Organization is $0 - $25 per Individual

3. Vaccinations Prevented 5.3 Million Illnesses
4. Vaccinations Prevented 2.6 Million Medical Visits
5. Vaccinations Prevented 85,000 Hospitalizations

The Epidemic / Pandemic BCP Team

Unique events such as epidemics and pandemics require unique solutions to mitigate risks and reduce potential impacts. Just as the Epidemic / Pandemic BCP is unlike any other component of the entire organizational BCP, the team chosen to manage this plan must be unique as well. The team composition will change from entity to entity based on organizational structure and need, but BCP managers should attempt to include internal or external team members with medical training whenever possible.

Listed below are five departments from which BCP team members should be selected, as these departments will have a significant impact before, during, and after an epidemic or pandemic.

1. Human Resource Management
2. Information Technology Management
3. Facilities Management
4. Logistics Management
5. Sales and Operations Management

The Epidemic / Pandemic Risk Assessment

The risk assessment for an epidemic or pandemic is based on a fluctuating business impact analysis and will vary based on the organizational unit's responsibility and mission. The business impact analysis and subsequent risk assessment will change from entity to entity based on organizational structure and need.

Listed below are six areas of consideration for the BCP manager conducting this type of assessment.

1. Employee-to-Employee Contact (Internal)
2. Employee-to-Customer Contact (External Sales)
3. Contact with Infected Items (Rent and Return)
4. Contact from Travel (Sales and Operations)
5. Impact on Raw Materials and Supplies (Upstream)
6. Impact on Customer Demand (Downstream)

Human Resource Management

Human Resource and Information Technology Management each have responsibilities that provide them with unique roles during the execution of an epidemic or pandemic BCP strategy. They are both in a position to recognize the organizational policies that are most likely to be impacted by such an event, and they are both in a position to implement risk mitigation solutions to lessen impact.

In short, Human Resource Management can create the flexibility in employee policies required to minimize impact on the operation, and Information Technology Management can create the distance between staff required to minimize the spread of the influenza. The extent to which organizational policies can be modified will change from entity to entity based on organizational structure and need. Listed below are six policy areas the BCP manager should consider when developing the Epidemic / Pandemic BCP.

1. Review and Modify the Virtual Worker Policy
2. Review and Modify the Attendance Policy
3. Identify Trained Substitutes for Key Personnel
4. Review and Modify the Company Travel Policy
5. Consider Rotating, Alternate, and Flexible Shifts
6. Consider a Free Immunization Policy (Voluntary)

Remote Worker Security Risks

The best way to prevent the spread of any disease is by creating distance between people. An organization can accomplish this in many ways, but the implementation of a remote worker policy is the most common solution. If a BCP manager is creating a plan for an organization with a limited remote capability, they must understand the risks associated with allowing staff this capability if they have not had it previously. Listed below are statistics from a *Cisco/InsightExpress* (https://www.cisco.com) survey of staff using organizational laptops while away from the office.

1. 40% Admitted to Online Shopping
2. 21% Admitted to Sharing Their Device
3. 10% Admitted to Risky Wireless Behavior
4. 50% Admitted to Using Personal Devices on the Network
5. 38% Admitted to Opening ALL Email Attachments
6. 4% Admitted to Loss or Theft of a Company Device

The Epidemic and Pandemic Communication Plan

An effective communication plan allows an organization to provide accurate information to staff well in advance of the disaster and test current mass communication methods to ensure that contact information is correct. Listed below are six common communication methods the BCP manager should consider when developing the epidemic / pandemic communication plan.

1. Posters and Flyers
2. Team Meetings
3. SMS Notifications
4. The Organization's Website
5. Toll-Free Hotlines
6. CDC Instructional Videos (Posted on the Website)

Common Area Sanitation Plan

Common area sanitation plans help to reduce the spread of disease and bring precautionary awareness to the forefront of peoples' minds. These tasks must be assigned; do not expect people to volunteer. Listed below are six common areas and fixtures the BCP manager should consider when developing the sanitation plan.

1. Doorknobs and Push Plates
2. Banister Rails
3. Light Switches
4. Lunchroom, Breakroom, and Meeting Areas
5. Vending Machines
6. Shared Workstations and Equipment

Post-Epidemic / -Pandemic Considerations

The scope of post-incident actions will change from entity to entity based on the extent of the impact to the organizational structure. BCP managers must maintain vigilance as epidemic/pandemic cases decline and utilize continuous communication until the event is officially declared over. Listed below are six actions the BCP manager should consider when the crisis has officially passed.

1. Make an Official Announcement that the BCP is Closed
2. Review the Effectiveness of the Plan
3. Identify the Impact on Employees and Families
4. Identify the Impact on Sales, Services, and Products
5. Identify the Impact on Suppliers, Customers, and Logistics
6. Provide Formal Recognition for People Involved

Preparing for Epidemics and Pandemics

KNOWLEDGE ASSESSMENT QUESTIONS

The following knowledge assessment questions are presented in true / false, multiple choice, and fill-in-the-blank formats. The correct answers are provided in an Answer Key at the end of Chapter 14. These questions may or may not be presented on quizzes and/or tests given by the instructor of this course.

Knowledge Assessment Questions

1) "*A widespread occurrence of a disease in a community that spreads quickly and affects many individuals at the same time*" is the definition of a(n) _____.

A. Epidemic

B. Influenza Outbreak

C. Pandemic

D. Common Cold

2) Which of the choices listed below is a distinctive characteristic of an epidemic that distinguishes it from a pandemic?

A. Occurs Rarely

B. Unpredictable Patterns

C. High Risk to Healthy Patients

D. Some Immunity Exists

3) The World Health Organization has divided pandemics into six phases. The _____ phase is typically the point at which a disaster is declared.

A. Animal to Human

B. Country to Country (Region)

C. Human to Human (Community)

D. Country to Country (Global)

4) Which of the choices listed below is a characteristic that is common to both influenza epidemics and pandemics?

A. A Contagious Immune System Illness

B. A Bacteria-Based Infection

C. Most Contagious in the First 1-4 Days

D. An Infection Radius up to 12 Feet

5) "*An epidemic that becomes very widespread and affects a whole region, a continent, or the world due to a susceptible population and causes a high degree of mortality*" is the definition of a(n) _____.

A. Epidemic

B. Influenza Outbreak

C. Pandemic

D. Common Cold

Knowledge Assessment Questions

6) Which of the choices listed below is a distinctive characteristic of a pandemic that distinguishes it from an epidemic?

A. Lower Risk to Healthy Patients

B. No Vaccines Initially Available

C. Predictable Patterns

D. Some Immunity Exists

7) When writing an epidemic / pandemic plan into the Business Continuity Plan, it is important to consider the _____ of the organizational staff.

A. Health Benefits

B. Job Titles

C. Transportation Needs

D. Family Members

8) Which of the choices listed below is an organizational policy which should be reviewed by Human Resources when creating the epidemic / pandemic plan?

A. Review the Attendance Policy

B. Identify Temporary Workers

C. Consider Mandatory Immunizations

D. Review Fixed and Permanent Shifts

9) The epidemic / pandemic _____ is based on a fluctuating business impact analysis and will vary based on unit responsibility.

A. Risk Analysis

B. Severity Rating

C. Vulnerability Assessment

D. Immunization Program

10) Which of the choices listed below is a characteristic of potentially vulnerable populations which should be considered when creating the epidemic / pandemic plan?

A. Children Over the Age of 6

B. People with Documented Allergies

C. People with Body Mass Index ≤ 20

D. Adults Over the Age of 65

True / False Questions

11) Maintain vigilance as epidemic / pandemic cases decline and utilize continuous communication until the event is officially declared over.

1. True
2. False

12) Common area sanitation tasks should be considered by BCP management. Request volunteers and trust those who do to manage their tasks properly.

1. True
2. False

13) Provide accurate information to organizational staff well in advance of the disaster event and test SMS messages to verify current contact numbers.

1. True
2. False

14) Public Relations and Senior Leadership are the core of the epidemic / pandemic plan and can identify company policies impacted by risk mitigation efforts.

1. True
2. False

15) Unique events require unique solutions; include organizational staff with medical and emergency services training in the epidemic / pandemic plan if possible.

1. True
2. False

16) When writing an epidemic / pandemic plan into the Business Continuity Plan, it is important to consider the vacation schedules of the organizational staff.

1. True
2. False

17) Despite the distinction of definitions between epidemics and pandemics, an epidemic can be viewed as a localized pandemic and treated as such.

1. True
2. False

True / False Questions

18) The Centers for Disease Control and Prevention has divided pandemics into nine phases. The phase number will dictate when a disaster is declared.

1. True
2. False

19) The epidemic / pandemic risk assessment is based on a fluctuating business impact analysis and will vary based on unit responsibility.

1. True
2. False

20) According to World Health Organization statistics, annual and seasonal epidemics worldwide typically can be expected to occur in the summer months.

1. True
2. False

The Role of Cloud Computing in Disaster Preparedness

KEY KNOWLEDGE POINTS

The Benefits of Cloud-Based Recovery Solutions
Cloud Computing Characteristics
Defining the Existing Operational State
Benefits of Cloud Computing
Cloud Computing Service Models
Cloud Computing Deployment Models

The Definition of Cloud Computing

NIST SP 800-145:

> Cloud computing is a model for enabling ubiquitous, convenient, on-demand network access to a shared pool of configurable computing resources (e.g., networks, servers, storage, applications, and services) that can be rapidly provisioned and released with minimal management effort or service provider interaction. This cloud model is composed of five essential characteristics, three service models, and four deployment models.

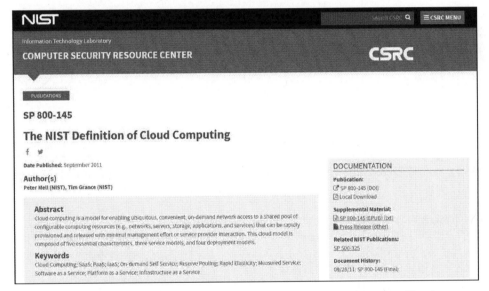

https://csrc.nist.gov/publications/detail/sp/800-145/final

Cloud Computing Characteristics

Despite many functions and configurations, five characteristics are accepted as part of the cloud computing definition.

1. On-Demand Self-Service
2. Broad Network Access
3. Resource Pooling
4. Rapid Elasticity
5. Measured Service (Metering)

Defining the Existing Operational State

Although cloud platforms are gaining in popularity as solutions for BCP/DRP managers, not every organization needs to migrate their data. The purpose of identifying the organizations' existing operational state is to determine its needs, not the solutions for those needs.

Analysis and assessment methods will change from entity to entity based on organizational structure and needs. Listed below are seven actions typically taken by BCP managers when defining the existing operational state of an organization.

1. Interview Senior Leadership
2. Interview Unit Managers
3. Interview Internal End Users of Systems
4. Interview External Customers
5. Collect and Analyze Marketing Data
6. Collect and Analyze Network Traffic
7. Identify Regulatory and Compliance Requirements

Reducing Infrastructure Expenses

Migrating data to the cloud as part of the BCP can be an attractive option for many organizations. Technical resources tend to be either underutilized which wastes money, or overutilized to the point of potential failure.

Cloud platforms minimize this concern by providing metered services—"pay as you go"—that allow clients to only pay for the services they need. The cloud can also serve as a backup during peak periods to support existing internal legacy systems, a term referred to as "cloud bursting."

Reducing Personnel and Payroll Expenses

Organizations may also take advantage of cloud services to reduce personnel and payroll expenses associated with IT functions. Managing data and its required infrastructure is neither a core function of most organizations, nor is it a profit center for the business process itself. Managing data requires specialized skills.

It is difficult to recruit and retain qualified IT personnel and they tend to be disproportionately more expensive than non-technical staff. Cloud platforms minimize this concern by reducing the number of data management staff needed by the organization and including the costs of data management as part of the contract.

Sharing Regulatory and Compliance Costs

Even in a cloud environment, the BCP manager must be aware that organizations cannot transfer the risk or liability associated with disclosure of *Personally Identifiable Information* (PII). However, the cloud is structured such that regulatory compliance packages can be included when required by specific industries and applied to the customer account.

These packages are negotiated into the service contract and remove the organizations' need to manage individual controls, reducing both organizational expenses and effort.

Cloud Computing Service Models

Cloud services are typically offered in three standard models based on provider capability and customer needs. The types of models selected will change from entity to entity based on organizational structure and BCP goals.

Listed below are the three standard cloud models and a few examples of their corresponding functions.

1. **IaaS** (Infrastructure as a Service): In this model, the provider is responsible for providing the hardware and the infrastructure to manage their data needs.

2. **PaaS** (Platform as a Service): In this model, the provider is responsible for providing IaaS capabilities, but also provides the client with numerous varieties of operating systems as needed.

3. **SaaS** (Software as a Service): In the model, the provider is responsible for providing IaaS and PaaS capabilities, but also provides the client with applications such as accounting and email as needed.

Infrastructure as a Service (IaaS)

In the *IaaS* model, the cloud service provider contracts access to its infrastructure and is fully responsible for its administration. This model is typically utilized for clients with limited needs, such as archiving data for BCP/DRP purposes. Utilization of this model will change from entity to entity based on organizational structure and needs.

The cloud customer provides the operating systems, required applications, and is responsible for the maintenance of both. This tends to increase security for the clients' data, as the party controlling the O/S also controls security.

Platform as a Service (PaaS)

In the *PaaS* model, the cloud service provider offers not only the IaaS capability but provides the customer with operating systems as well. The provider can offer clients multiple operating systems simultaneously which can be useful for software development that must be tested in isolated (non-production) environments.

Utilization of this model will change from entity to entity based on organizational structure and software development needs. These operating systems are typically hardened, maintained by the provider, and allow developers to conduct compatibility testing for their software with multiple O/S's.

Software as a Service (SaaS)

In the *SaaS* model, the customer receives all the underlying IaaS and PaaS capabilities, as well as end user applications that are needed by the organization. The cloud provider is responsible for the administration of the infrastructure, operating systems, and all applications. This is a hosted full production environment in which the client is only required to upload and utilize data.

Utilization of this model will change from entity to entity based on organizational structure and the applications required for daily operations. Listed below are a few typical uses of the SaaS model.

1. Google Docs, MS Office 365
2. *Customer Relationship Management* (CRM) Software
3. Accounting Software

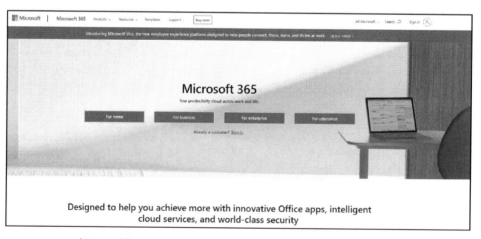

Designed to help you achieve more with innovative Office apps, intelligent cloud services, and world-class security

https://www.microsoft.com/en-us/microsoft-365

Public Cloud

When discussing cloud service providers and cloud capabilities, the *public cloud* tends to be the solution to which is being referred most often. All cloud resources are owned by the provider and offered to the general public through lease or contract agreements.

The decision to utilize public cloud services will change from entity to entity based on existing infrastructure and organizational needs. Listed below are six examples of public cloud providers a BCP manager will most likely encounter when researching cloud-based solutions.

1. Microsoft Azure
2. *Amazon Web Services* (AWS)
3. Rackspace Managed Cloud
4. Google Cloud Platform (G-Suite)
5. IBM Cloud
6. Oracle Cloud
7. Alibaba Cloud

Private Cloud

Private clouds are owned and operated by individual organizations for the specific use of their staff, customers, and vendors. They are based on standard IT legacy environments of datacenters and all supporting infrastructure. Staff, customers, and vendors can connect to an organizations' private cloud through the Internet with a web browser using a remote access capability.

The configuration of the private cloud will change from entity to entity based on existing infrastructure and services needed. Formerly referred to as "intranets," private clouds are typically used for shared storage and internal resources such as a hosted SharePoint solution. Listed below are five use cases for private cloud a BCP manager should be aware of when considering this option.

1. Potential Cost Savings
2. Agile Development Environment
3. Deployment of Production Workloads
4. Enhanced Flexibility and Transparency
5. Control over Security and Compliance

Community Cloud

Community clouds provide infrastructure and functionality that is owned and operated by affinity groups and similar organizations. Segments are owned and maintained by individual organizations, while the responsibility of joint tasks and functions are assumed by the group as a whole. Examples of community cloud solutions can be found in online gaming communities. Listed below are examples of how constituent members of a community cloud such as PlayStation would interface and interact at various levels.

1. PlayStation: The Network Umbrella Unit
2. Sony: Hosts the Network IAM Functionality
3. Game Vendors: Host DRM Servers
4. End Users: Conduct Testing and Processing

Hybrid Cloud

Hybrid clouds contain elements of public, private, and community cloud models to varying degrees based on customer needs. In this model, private cloud resources are retained—such as the organizations' legacy infrastructure—as well as legacy production environments for operations. Applications and software are retained for specific use by staff, customers, and vendors and accessed remotely.

The organization leases space on the public cloud (typically a PaaS model) for software development. This hybrid model protects the legacy production environment and exposes associated systems to less risk of harm.

The Role of Cloud Computing in Disaster Preparedness

KNOWLEDGE ASSESSMENT QUESTIONS

The following knowledge assessment questions are presented in true / false, multiple choice, and fill-in-the-blank formats. The correct answers are provided in an Answer Key at the end of Chapter 14. These questions may or may not be presented on quizzes and/or tests given by the instructor of this course.

Knowledge Assessment Questions

1) Creating a dedicated _____ may prove to be a costly option in terms of the infrastructure, technology, personnel, and safety considerations required.

A. Risk Assessment

B. Continuity Plan

C. Data Backup

D. Recovery Site

2) Which of the choices listed below is a distinctive characteristic of the private cloud that distinguishes it from other cloud provider deployment models?

A. Formerly Termed "Intranets"

B. Resources are Leased to Customers

C. Segments are Owned by Organizations

D. Resources are Offered to the Public

3) *"A category of cloud computing services that allows customers to provision, instantiate, run, and manage a modular bundle comprising a computing platform and one or more applications"* is the definition of _____.

A. Software as a Service (SaaS)

B. Cloud Computing

C. Platform as a Service (PaaS)

D. Infrastructure as a Service (IaaS)

4) Which of the choices listed below can be considered a benefit of utilizing a cloud-based solution in business continuity planning?

A. Enables Staff to Work from One Location

B. Easily Implemented with High Reliability

C. Models are Tailored to the Providers' Needs

D. Multiple Locations Increase Redundancy

5) Even in a cloud environment, organizations cannot transfer the liability or responsibility associated with _____ of Personally Identifiable Information.

A. Storage

B. Disclosure

C. Aggregation

D. Characteristics

Knowledge Assessment Questions

6) Which of the choices listed below is a distinctive characteristic of the community cloud that distinguishes it from other cloud provider deployment models?

A. A Standard IT Legacy Environment

B. Resources are Owned by the Provider

C. Owners Perform Joint Tasks and Functions

D. Formerly Termed "Intranets"

7) *"A category of cloud computing services that allows customers to license software and a delivery model in which software is licensed on a subscription basis and is centrally hosted"* is the definition of _____.

A. Software as a Service (SaaS)

B. Cloud Computing

C. Platform as a Service (PaaS)

D. Infrastructure as a Service (IaaS)

8) Which of the choices listed below is a distinctive characteristic of Infrastructure as a Service (IaaS) that distinguishes it from other cloud provider service models?

A. The Provider Installs the Software

B. The Hardware is Controlled by the Customer

C. The Provider Installs the Operating Systems

D. Cloud Service is Utilized for Limited Needs

9) When discussing cloud service providers and cloud capabilities, the _____ cloud tends to be the solution to which is being referred most frequently.

A. Private

B. Community

C. Hybrid

D. Public

10) Which of the choices listed below is one of the five characteristics which are accepted as part of the NIST SP 800-145 cloud computing definition?

A. On-Demand Self-Service

B. Broad Internet Access

C. Incremental Elasticity

D. Measured Reporting

True / False Questions

11) Service Level Agreements contain elements of public, private, and community cloud models to varying degrees based on customer needs.

1. True
2. False

12) Community clouds provide infrastructure and functionality that is owned and operated by affinity groups and similar organizations.

1. True
2. False

13) In the IaaS model, the cloud service provider contracts access to its infrastructure and the customer is fully responsible for its administration.

1. True
2. False

14) Cloud services are typically offered in three models based on the capability of the provider and the needs of the customer.

1. True
2. False

15) Even in a cloud environment, organizations cannot transfer the liability or responsibility associated with disclosure of audit and compliance information.

1. True
2. False

16) Managing data and its required infrastructure is neither a core function of most organizations nor is it a profit center to the organizational process.

1. True
2. False

17) Organizations tend to underutilize a technical resource (potential failure) or overutilize a technical resource (wasted money).

1. True
2. False

True / False Questions

18) The purpose of identifying the organizations' operational state is to determine its needs, not the solutions for those needs.

1. True
2. False

19) Despite many functions and configurations, there are seven characteristics which are accepted as part of the NIST SP 800-145 cloud computing definition.

1. True
2. False

20) Creating a dedicated recovery site may prove to be a costly option in terms of the infrastructure, technology, personnel, and safety considerations required.

1. True
2. False

Cloud Security Risks and Threat Vectors

KEY KNOWLEDGE POINTS

Cloud Platform Risks
Private and Public Cloud Risks and Threats
Vendor Lock-Out
IaaS / PaaS / SaaS Risks and Threats
Virtualization Risks and Threats
Cloud-Specific BIA Risks

Cloud Platform Risks

Because the cloud customer and provider will each process data, they will share responsibilities and risks associated with the data. In the event of unauthorized use or disclosure of PII, the provider may be financially liable for damages if that responsibility is codified in the service contract. Although the customer may recover the financial losses stipulated in the contract, the customer assumes all responsibility and ownership of criminal and civil liabilities.

In addition, if the customer is protected by the providers' acceptance of financial responsibility, legal repercussions are not the only negative impact on the organization. Listed below are five impacts likely to be experienced by the customer while the litigation process works its way through the legal system.

1. Negative Publicity
2. Loss of Clientele Faith
3. Decrease in Market Share
4. Decrease in Share Value (Publicly Held Organizations)
5. Increase in Insurance Costs

Private Cloud Risks

A private cloud configuration is a legacy configuration of a datacenter with distributed computing capabilities. There are no risks unique to private clouds that do not impact other platforms. The benefit of a private cloud as relating to risk is the organization retains complete control over the cloud environment, making any response much timelier and more efficient.

Listed below are five categories of risk typically considered by organizations utilizing a private cloud solution.

1. Personnel Threats
2. Natural Disasters
3. External Threats
4. Regulatory and Compliance Risks
5. Malware Risks (Internal and External)

Community Cloud Risks

In a community cloud configuration, resources are allocated, shared, and dispersed among affinity groups. This provides the members of the platform several benefits, but each of those benefits presents a risk to the cloud customer. The community cloud has shared ownership; if one node becomes unavailable the others will continue to operate.

However, each node in a shared environment also presents itself as a unique point of entry.

Costs are shared and distributed, but the responsibility for access and overall control of the environment is shared as well. This also removes the need for centralized administration but at the same time denies members the reliability of centralized standards.

Public Cloud Risks

In a public cloud configuration, a company offers cloud services to any entity wanting to become a cloud customer. While this option poses the same risks as those impacting private and community clouds, it also exposes the customer to risks unique to this specific environment. Listed below are six risks typically experienced by customers that are unique to public cloud platforms.

1. The Customer Loses Control of Data
2. The Customer Loses Oversight of Operations
3. The Customer Loses all Audit Ability of Data and Systems
4. The Customer Loses the Ability to Enforce Policy
5. The Customer Risks the Potential of Vendor Lock-In
6. The Customer Risks the Potential of Vendor Lock-Out

Vendor Lock-Out

Vendor lock-out can be caused when the provider goes out of business, is acquired by another interest, or ceases operation for any reason. This risk is rare but still must be considered. Listed below are five public cloud provider characteristics the BCP manager should consider when assessing potential risk.

1. The Providers' Longevity
2. The Providers' Core Competency
3. The Providers' Jurisdictional Suitability
4. The Providers' Supply Chain Dependencies
5. The Legislative Environment for the Provider and Client

Risks: IaaS (Infrastructure as a Service)

In the IaaS (Infrastructure as a Service) model, the customer will have the most control over its resources. Only the infrastructure (hardware) is maintained and administered by the provider; the customer manages all other aspects of the cloud operation. This level of control presents the organization with potential risks.

The risks associated with the IaaS model will change from entity to entity based on organizational structure and staffing.

Listed below are six potential areas of risk that should be considered by the BCP manager conducting a risk assessment or analysis of this model.

1. Personnel Threats
2. External Threats
3. Lack of Required Skill Sets (Organizational Staff)
4. All Access will be by Remote Connection
5. Responsibility for all Operational and Security Functions
6. Lack of Sufficient Personnel to Manage Solution

Risks: PaaS (Platform as a Service)

In the PaaS (Platform as a Service) model, the customer will have all the risks associated with the IaaS model as well as additional risks specific to the PaaS model. The risks associated with the PaaS model will change from entity to entity based on organizational structure and staffing.

Listed below are three potential areas of risk that should be considered by the BCP manager conducting a risk assessment or analysis of this model.

1. **Interoperability Issues**: The PaaS model requires the provider to provide, administer and update the operating systems required by the customer. Issues may arise when unannounced version and/or patch management actions are initiated by the provider but unknown to the customer. This may impact critical software functionality and *Application Programming Interfaces* (API's).

2. **Persistent Backdoors**: A benefit of the PaaS model is that it allows software developers (DevOps) to create and test software in non-production environments. In many cases developers install persistent backdoors in the software to ease the burden of making iterative changes anywhere in the process. If this software is transferred to a production environment without removing these backdoors, today's DevOps project may become tomorrow's zero-day exploit.

3. **Virtualization Risks**: Most virtualization solutions utilized by cloud customers reside on resources shared by other cloud customers. Potential risks such as information bleed and side-channel attacks can occur if virtualization security controls are not effective. Unfortunately for the cloud customer, they have no ability to attenuate these potential risks, and must rely entirely on the provider to ensure these risks are recognized and mitigated.

Risks: SaaS (Software as a Service)

In the SaaS (Software as a Service) model, the customer will have all the IaaS and PaaS risks listed above as well as risks specific to the SaaS model. In this model the customer has no control over the cloud environment and must be especially vigilant when conducting operations.

The risks associated with the SaaS model will change from entity to entity based on organizational structure and software utilization.

Listed below are five potential areas of risk that should be considered by the BCP manager conducting a risk assessment or analysis of the SaaS model.

1. Customer Data Stored in Proprietary Formats
2. Virtualization Risks Cannot be Attenuated by the Customer
3. Web Application Security Issues
4. Customer Access is Required Solely via Web Browser
5. Potential Weaknesses in API's Disrupting Applications

Virtualization Risks

As stated in the previous sections, many potential threats are posed by virtualization in public cloud environments that can only be attenuated via the use of controls that can only be implemented by the cloud service provider. An attack on the hypervisor is the most serious of those concerns.

The risks associated with the virtualization will change from entity to entity based on organizational structure and architecture.

Malicious actors tend to prefer Type II hypervisors—those that reside on top of an operating system as opposed to residing on bare metal (a Type I hypervisor)—because the associated operating system provides a much greater attack surface. The Type II configuration is typically used by cloud providers for their customers.

Listed below are five potential areas of risk that should be considered by the BCP manager conducting a risk assessment or analysis of a cloud virtualization implementation.

1. Guest Escape
2. Host Escape (Rare, but Possible)
3. Information Bleed (Shared Resources)
4. Identification of Process-Specific Information
5. Data Seizure by Law Enforcement (Collateral Damage)

Private Cloud Threats

Although many threats to cloud computing are the same as those faced in legacy operations, they may manifest in novel ways and present new risks. The threats associated with private cloud solutions will change from entity to entity based on organizational policies and system architecture.

Listed below are seven potential threats that should be considered by the BCP manager conducting a risk assessment or analysis of a private cloud platform.

1. Internal Threats (and External Contractors)
2. External Attackers
3. Man-in-the-Middle Attacks
4. Increased Exposure via Remote Access Capability
5. Social Engineering
6. Theft and/or Loss of Devices (BYOD Environments)
7. Regulatory and Compliance Violations

Public Cloud Threats

The public cloud not only includes all threats associated with private clouds but includes threats totally outside the customers' ability to defend against and control. These threats can cause more damage in cloud environments than they would in legacy systems.

Additionally, two sets of governance and guidelines—the providers' and the customers'—hinder the process of securing permission to resolve certain issues.

To perform their work effectively, technicians may be inclined to conduct unapproved operations at a higher rate under these restrictions. Listed below are three potential threats that should be considered by the BCP manager conducting a risk assessment or analysis of a public cloud platform even if they cannot be controlled.

1. Rogue Administrators
2. Escalation of Privileges
3. Contractual Failure (Unlikely, but Possible)

Cloud-Specific BIA Risks

Unlike a legacy environment, a customer conducting operations in the cloud will not be able to conduct local computing without the provider (unless the cloud platform is a hybrid). The BIA must also consider a new set of upstream and downstream dependencies for the cloud service as well. The ease of data distribution and storage across diverse geographical locations presents an additional set of regulatory and compliance risks.

Additionally, the inclusion of internal personnel and increased remote access elevate the threat of data breaches and unauthorized disclosures of PII. Finally, the risks of both vendor lock-in and vendor lock-out must also be considered when assessing cloud computing for BCP needs.

CHAPTER 09

Cloud Security Risks and Threat Vectors

KNOWLEDGE ASSESSMENT QUESTIONS

The following knowledge assessment questions are presented in true / false, multiple choice, and fill-in-the-blank formats. The correct answers are provided in an Answer Key at the end of Chapter 14. These questions may or may not be presented on quizzes and/or tests given by the instructor of this course.

Knowledge Assessment Questions

1) Because the cloud customer and provider will each process data, they will share responsibilities and risks associated with the _____.

A. Cost

B. Location

C. Contract

D. Data

2) Which of the choices listed below is a risk associated with Personally Identifiable Information (PII) when utilizing a cloud provider platform?

A. The Provider is Financially Liable

B. The Customer cannot Recover Damages

C. The Provider Owns the Civil Liability

D. The Provider Owns the Criminal Liability

3) Although the customer is protected by the providers' acceptance of financial responsibility, legal repercussions are not the only _____ to expect.

A. Positive Benefit

B. Contract Stipulation

C. Negative Impact

D. Local Regulations

4) Which of the choices listed below represents a potential risk to customers when operating in a community cloud platform environment?

A. Resiliency Through Shared Ownership

B. No Reliability of Centralized Standards

C. Centralized Administration is not Needed

D. Shared and Distributed Costs

5) _____ can be caused when the provider goes out of business, is acquired by another interest, or ceases operation for any reason.

A. Increased Usage

B. Vendor Lock-Out

C. Contract Negotiations

D. Vendor Lock-In

Knowledge Assessment Questions

6) Which of the choices listed below is a factor which should be considered when researching a cloud provider to avoid the risk of vendor lockout?

A. Business Affiliations

B. Marketing Capabilities

C. Core Competencies

D. Number of Employees

7) Many potential threats posed by _____ require attenuation via the use of controls that can only be implemented by the cloud service provider.

A. Virtualization

B. External Actors

C. Natural Disasters

D. Underutilization

8) Which of the choices listed below represents a potential risk to customers when utilizing the platform of any cloud service provider?

A. Fluctuations in Share Value

B. Loss of Provider Faith and Loyalty

C. Potential Decrease in Insurance Costs

D. Damage to the Brand Image

9) Unlike a traditional legacy environment, a customer conducting operations in the cloud will not be able to conduct _____ without the provider.

A. Productivity Assessments

B. Business Processes

C. Media Campaigns

D. Local Computing

10) Which of the choices listed below is a risk factor which should be considered when conducting a cloud-specific Business Impact Analysis?

A. The Potential of Regulatory Failure

B. Decreased Ease of Data Distribution

C. The Customers' Internal Personnel

D. No New Dependencies are Created

True / False Questions

11) Because the cloud customer and cloud provider will each process data, they will share responsibilities and risks associated with market conditions.
1. True
2. False

12) A private cloud configuration is a traditional legacy configuration of a datacenter with a variety of distributed computing capabilities.
1. True
2. False

13) In a hybrid cloud configuration, resources are allocated, shared, and dispersed among affinity groups.
1. True
2. False

14) In a public cloud configuration, a company offers cloud services to any entity wanting to become a cloud customer and is willing to accept provider terms.
1. True
2. False

15) Vendor lock-in can be caused when the provider goes out of business, is acquired by another interest, or ceases operation for any reason.
1. True
2. False

16) In the IaaS model (Infrastructure as a Service), the customer will have the most control over its resources and responsibility for asset oversight.
1. True
2. False

17) In the PaaS model (Platform as a Service), the customer will have all SaaS risks as well as those associated with web application responsibility and oversight.
1. True
2. False

True / False Questions

18) In the SaaS model (Software as a Service), the customer will have all IaaS and PaaS risks as well as those associated with applications and access.

1. True
2. False

19) Many potential threats posed by virtualization require attenuation via the use of controls that can only be implemented by the cloud customers' security staff.

1. True
2. False

20) The public cloud not only includes all threats associated with private clouds but also includes threats totally outside the providers' knowledge and control.

1. True
2. False

Cloud Data Storage, Security, and Administration

KEY KNOWLEDGE POINTS

The Cloud Data Life Cycle (CDLC)
Volume and Object-Based Storage
The Content Delivery Network (CDN)
Foundations of Managed Cloud Services
Shared Responsibilities by Service Type
Lack of Physical Access and Auditing

The Cloud Data Life Cycle (CDLC)

Data stored in the cloud tends to have the same needs as data stored in legacy systems and should be treated in the same manner. Listed below are the six stages of the cloud data life cycle that demonstrate its similarity to the data life cycle in legacy systems.

1. *Create* Data
2. *Store* Data
3. *Use* Data
4. *Share* Data
5. *Archive* Data
6. *Delete* Data

CDLC: Create Data

Users can create data by accessing the cloud remotely or from within the cloud datacenter where the data resides. The means by which users create data will change from entity to entity based on organizational structure and policy. Whether data is created remotely or from within the cloud environment, the practices of information security will apply. Listed below are four principles typically implemented by users creating data in the cloud.

1. Encrypt the Data Prior to Uploading
2. Use a Cryptosystem with a High Work Factor
3. Use a Cryptosystem Listed on FIPS 140-2
4. Secure the Upload Connection (VPN, IPSec)

CDLC: Use Data

Operations within a cloud environment will require remote access with secured connections (typically an encrypted tunnel) utilized in such a way that the cloud provider cannot access raw client data. Users relying on remote access must be aware of potential risks and trained in the use of the technology needed to mitigate those risks. Users must also be informed of any and all VPN, DRM, and/or DLP requirements when interfacing with data. Leadership must be made aware that logging and audit trails are crucial in this environment, and strong protections using virtualization are strongly encouraged.

CDLC: Share Data

Global collaboration is a powerful capability of the cloud, but the risks associated with that power are global as well. The fact that cloud users can be anywhere means threats can be anywhere as well. The means by which users share data will change from entity to entity based on organizational policy and location. The ability to share data in the cloud may be restricted by jurisdictional law, or by regulatory mandates. Listed below are three practices typically implemented when sharing data in cloud environments.

1. Encrypt all Files and Communications
2. Utilize Digital Rights Management Solutions
3. Utilize Export and Import Controls for Technology

CDLC: Archive Data

Archived data is stored for long periods of time, and these longer timeframes will tend to require special security considerations. When data is archived for BCP/DRP purposes cryptography is an essential component, key management is crucial, and the physical security of the data is equally as important. The means by which archived data is stored will change from entity to entity based on organizational policy and location. Listed below are four questions a BCP manager should ask when archiving data in the cloud.

1. Where is the Data Stored?
2. What Format is Used for Data Storage?
3. Who is Staffing the Storage Location?
4. What Data Transfer Procedures Exist?

Cloud Architecture: Volume Storage

There are two primary data storage methods associated with cloud architecture: *volume storage* and *object-based storage*. With the *volume storage* option, cloud customers are assigned storage space that is typically attached to a virtual machine. It uses allocated blocks for file-based storage that share the same hierarchy and structure as that found in legacy systems. Listed below are characteristics typically associated with volume storage which should be known before considering this storage method.

1. The Allocated Volume can Contain Anything
2. It has Higher Flexibility and Performance
3. It has More Administrative Overhead
4. This Method is Typically Used with IaaS Solutions

Cloud Architecture: Object-Based Storage

With an *object-based storage* solution, data is stored individually as objects as opposed to files or blocks. The data stored as objects includes not only production content, but extensive metadata for the content as well. The stored objects share a similarity with database objects as they are each assigned a unique address identifier. Listed below are four characteristics typically associated with object-based storage which should be known before considering this storage method.

1. Allows for Marking, Labels and Classification
2. Enhances Data Indexing Capabilities
3. Enhances Data Policy Enforcement
4. Allows for Centralized Data Management

The Content Delivery Network (CDN)

A *Content Delivery Network* (CDN) is a form of data caching typically located near geophysical locations of predicted high demand and use. Using this method data is not drawn from a datacenter; the CDN stores copies of media based on likelihood of user requests. This accommodates users at varying distances by allowing for increased bandwidth and delivery quality. The implementation of the CDN will change from entity to entity based on organizational policy and location, but it will typically be used for online multimedia streaming services.

Cloud Data Security: Encryption

Like its legacy environment counterpart, cloud computing solutions have a significant dependency on encryption to operate. In fact, no encryption would mean no cloud service. The client enterprise uses encryption to protect its data, and the cloud service provider uses encryption to isolate client data from other clients in a shared resource environment. Listed below are the four primary functions of encryption typically encountered in cloud environments.

1. Create Secure Remote Connections
2. Protect Data at Rest
3. Protect Data in Use
4. Protect Data in Transit

Cloud Data Security: Key Management

The methods and locations chosen to manage encrypted keys impacts the risk to data in several ways. The implementation of a key management will change from entity to entity based on organizational policy and need. In some instances, an organization may decide to outsource its key management program to a third party such as a *Cloud Access Security Broker* (CASB). In other cases, the key management may remain an internal responsibility. Regardless of the method chosen to manage encrypted keys there are five common areas of concern that must be considered and addressed.

1. Level of Protection Implemented
2. Creation of a Key Recovery Policy
3. Creation of a Key Distribution Policy
4. Creation of a Key Revocation Policy
5. Creation of a Key Escrow Policy

Obfuscation, Masking, and Anonymization

In certain instances, customers using cloud provider solutions may find it necessary to obscure data and use a representation of data instead. This is not considered to be a "best practice" within the information security field but there are times when these measures become necessary. The choice of using obfuscation, masking, and anonymization as a method for information security will change from entity to entity based on organizational policy and need.

This can be a viable option if supporting security controls are put in place such as "sandboxing" for test environments, least privilege enforcement within the network, and the establishment of a secure remote access policy for all users with access to the data. Listed below are four examples of obfuscation, masking, and anonymization typically implemented by organizations within their information security program.

1. Hashing (One-Way Function)
2. Randomization (Replacement of Data)
3. Masking (Hiding Data with Character Shuffling)
4. Shuffling (Randomization Using Production Data)

Security Information and Event Management (SIEM)

To better collect, manage, analyze, and display log data, a set of SIEM tools have been developed specifically for that purpose. SIEM is an approach to security management that combines SIM (security information management) and SEM (security event management) functions into one security management system. The underlying principles of every SIEM system is to aggregate relevant data from multiple sources, identify deviations from the norm and take appropriate action.

One of the primary drivers for SIEM solutions is that humans are not very efficient at long-term analysis of logs. When an individual spends too much time analyzing logs, the information tends to blend together and events may be overlooked. When someone spends too little time analyzing logs, a knowledge and experience gap is created that leads to the same result of missed events. The choice of using a SIEM solution as a method for information security will change from entity to entity based on organizational policy and need.

The choice to implement a SIEM solution is not entirely without risk: aggregated data is stored in one location making it vulnerable to compromise. Despite this risk, there are many benefits to the implementation of a SEIM solution, six of which are listed below.

1. Centralized Collection of Log Data
2. Enhanced Log Analysis Capabilities
3. Trend Detection in Large Datasets
4. Dashboarding (Management Display)
5. Automated Response Capabilities
6. Data Storage and Normalization

Data Loss Prevention: Egress Monitoring

An egress monitoring solution will examine data leaving the production environment and react based on preestablished rules and parameters. It is usually implemented as an additional layer of security for the purpose of Data Loss Prevention (DLP). The choice of using a DLP solution and the data to be monitored will change from entity to entity based on organizational policy and need. Listed below are five functions of egress monitoring that may offer an organization the benefit of an additional security control.

1. Prevents the Inadvertent or Malicious Disclosure of Data
2. Serves as a Policy Enforcement Mechanism
3. Provides Enhanced Monitoring Capabilities
4. Creates an Enhanced Capability for Regulatory Compliance
5. Establishes a Higher Level of Data Dissemination Control

Foundations of Managed Cloud Services

Some degree of adversarial relationship exists between the cloud customer and the provider because they have somewhat different goals. For these reasons, contracts and Service Level Agreements are crucial components of a stable relationship. The "gray areas" not formally defined by a contract are where this risk for both parties exists. Listed below are five examples of differences between customers and providers in the goals of their relationship.

1. Customer: Seeks to Maximize Operational Capabilities
2. Customer: Seeks to Maximize Data Security
3. Customer: Seeks to Minimize Operational Expenses
4. Provider: Seeks to Maximize Profits
5. Provider: Seeks to Minimize Service Labor for Customer

Provider Responsibility: Physical Plant

The provider is responsible for the administration of the physical plant. The physical plant of the datacenter will include the facility campus, all physical components, and the services that support them. Responsibilities for administration of the physical plant will change from provider to provider based on organizational structure and needs. Listed below are seven areas of provider responsibility typically associated with administration of the physical plant for the cloud customer.

1. Ensure the Use of Secure Hardware Components
2. Ensure the Use of TPM Standards for BIOS Firmware
3. Ensure the Use of a Managed Hardware Configuration
4. Establish Controls for Logging Events and Incidents
5. Establish Policies for Incidents, Forensics, and Attribution
6. Maintain all Components for Customer Needs
7. Ensure Configuration for Secure Remote Access

Provider Responsibility: Secure Logical Framework

In addition to securing the hardware components, the cloud provider must ensure that the logical elements are equally protected. The provider must take all reasonable steps to prevent data leakage and malicious aggregation of the customer data. The steps taken to secure the providers' logical framework will change from provider to provider based on organizational structure and capability. Listed below are five areas of provider responsibility typically associated with securing logical elements for the cloud customer.

1. Ensure the Use of Virtual Operating Systems
2. Ensure the Use of Virtualization Management Tools
3. Ensure Configuration Policy Enforcement
4. Ensure the Configuration of VM Elements is Secure
5. Ensure the Attenuation of Potential Risks

Provider Responsibility: Secure Networking

Secure networking often involves the same tactics and methods used in legacy environments, but with cloud-specific permutations. The steps taken to secure the customers' network will change from provider to provider based on organizational structure and needs. Listed below are seven categories of security controls typically implemented when securing networks for the cloud customer.

1. Firewalls
2. IDS / IPS

3. Honeypots
4. Use of Vulnerability Assessments
5. Systems to Protect Secure Communication
6. Use of Encryption and Strong Authentication
7. Use of Virtual Private Networks (VPN's)

Provider Responsibility: Mapping and Selection of Controls

The cloud provider must apply the proper security controls according to a customers' relevant regulatory frameworks and planned usage. A "one size fits all" mentality will never be an effective approach. The provider must create and make available published governance guides, and its subsequent security policies must be based on the published governance guides.

All security controls selected and implemented by the provider must be justified by recognized industry standards, such as those found in the CSA's *Cloud Controls Matrix* (cloudsecurityalliance.org). Additionally, the controls selected and implemented must be equally applicable to both the customers' needs and the providers' datacenter.

Shared Responsibilities by Service Type

By reviewing the type of service model utilized by the customer, the responsibilities can be shared and assigned to the appropriate party. Listed below are the three primary cloud service models and the level of shared responsibility typically associated with each.

1. **IaaS** (Infrastructure as a Service): The provider is responsible for the physical security of the plant and infrastructure, and the customer is responsible for all other functionality and services that follow.

2. **PaaS** (Platform as a Service): The provider is responsible for the physical security of the plant, the infrastructure, and the administration and security of all operating systems. The customer is responsible for all other functionality and services that follow.

3. **SaaS** (Software as a Service): The provider is responsible for the physical security of the plant, the infrastructure, the administration and security of all operating systems, and the administration and security of all software and applications utilized by the customer. The customer is responsible for assigning access and permissions to those services for vendors, staff, and customers.

Shared Responsibility: O/S Management

The operating system is a large attack surface and offers many potential vectors to malicious actors if not secured correctly. Even with a PaaS model in which the provider assumes responsibility for administration of the operating systems, there are several ways in which the provider and customer may collaborate to manage their security.

Listed below are seven areas in which the provider and customer can work together and manage the security of the O/S.

1. Removing Unnecessary Services
2. Closing Unused Ports
3. Installing Antimalware Agents
4. Limiting Administrator Access
5. Removing Default Accounts and Passwords
6. Enabling System Event and Incident Logging
7. Creating a Formal Approval Process for Deviations

Shared Responsibility: Data Access

In all cloud service models, the customer and their users will need to access and modify the data at various levels. This is another area in which cloud customers and providers can share responsibility and support each other. Listed below are methods by which the cloud customers and providers secure access to customer data.

1. **Customer Administration**: In this model, the customer has complete control and responsibility for data access in the cloud. This is always the case in IaaS (Infrastructure as a Service) models, as the customer assumes responsibility for the operating system and all associated security controls.

2. **Provider Administration**: In this model, the customer and the provider share the responsibility for data access. This requires good communication between both parties. The requests for access will be made through the provider, who then confirms with the customer (via automation) that the requests are legitimate.

3. **Third-Party Administration**: In this model, the customer and provider both work with a *cloud access security broker* (CASB) to manage data access. The CASB works with the customer and provider to verify user accounts and ensure the access requests are legitimate.

Shared Responsibility: Monitoring and Testing

An area where cloud providers and customers may find common ground in sharing responsibilities is in security monitoring and testing. The provider may allow the customer remote access to conduct testing in support of its own testing initiatives, but the access will be very restricted. The granting of limited access to a customer by a provider must be stipulated in a service contract.

Limiting the customer to specific aspects of the providers' system reduces the risk of significant harm to the infrastructure, and this of data disclosure within shared resources.

Challenge: Lack of Physical Access

The cloud provider will not have any reason to allow the customer any physical access to the facility containing customer data. This is both a challenging and beneficial situation for both parties. The cloud providers have every reason not to grant access to customers. The provider may work with thousands of customers, each of them having various levels of trust.

Additionally, the more a customer knows about the internal operations of the provider, the higher degree of risk exists of that knowledge being exploited. This lack of access can be beneficial insofar that it builds customer trust in the cloud providers' security. After all, every customer is denied access equally for security purposes.

The challenge of this lack of access is that the customer must rely on the providers' assertions of security controls without the ability to validate or verify those assertions firsthand.

Challenge: Lack of Auditing Ability

The cloud providers' unwillingness to allow customer access to the facility applies to the customers' auditors as well. The challenge exists with situations in which the cloud customer must demonstrate they are meeting regulatory and compliance requirements for data storage. This can also be frustrating for key stakeholders requiring audit results from the company.

In response to this challenge, cloud customers and providers agree through use of service contracts to rely on licensed and chartered auditors to conduct required audits that satisfy both parties.

These third parties publish "audit assurance statements" that are then made available to the customer and the provider. Different levels of audit assurance statements are made available for audiences with different needs.

Listed below are the three types of SOC (System and Organization Controls) reports typically encountered by individuals concerned with audit assurance.

1. **SOC 01**: This report falls completely within the domain of financial reporting and has no applicability to security in the cloud.

2. **SOC 02**: This report contains comprehensive details of the cloud provider audit and is never released to the public for security reasons.

3. **SOC 03**: This report is a brief audit assertion, based on the comprehensive audit, that is released to customers and the public. This is based on the *transitive model of trust*. If A (the provider) trusts B (the auditor), and if C (the customer) trusts B (the auditor), then by extension C (the customer) will trust A (the provider). That is the general theory as to why SOC 03 reports satisfy regulatory and compliance requirements.

Cloud Data Storage, Security, and Administration

KNOWLEDGE ASSESSMENT QUESTIONS

The following knowledge assessment questions are presented in true / false, multiple choice, and fill-in-the-blank formats. The correct answers are provided in an Answer Key at the end of Chapter 14. These questions may or may not be presented on quizzes and/or tests given by the instructor of this course.

Knowledge Assessment Questions

1) Encrypting data prior to uploading using a cryptosystem with a high work factor are components of step _____ of the Cloud Data Life Cycle (CDLC).

A. 01: Create the Data

B. 03: Use the Data

C. 04: Share the Data

D. 02: Store the Data

2) Which of the choices listed below is an option for obfuscation, masking, and anonymization that uses actual production data in its functionality?

A. Masking

B. Hashing

C. Randomization

D. Shuffling

3) With the _____ option, cloud customers using this architecture method are assigned storage space that is typically attached to a virtual machine.

A. Content Delivery Network

B. Volume Storage

C. Object-Based Storage

D. Private Cloud

4) Which of the choices listed below is a risk factor associated with Security Information and Event Management (SIEM) systems that should be considered prior to implementation?

A. Provides Trend Detection

B. Uses Dashboarding

C. All Logs are in One Location

D. Provides Automated Response

5) Adhering to Virtual Private Network, Digital Rights Management, and Data Loss Prevention requirements are components of step _____ of the Cloud Data Life Cycle (CDLC).

A. 01: Create the Data

B. 05: Archive the Data

C. 03: Use the Data

D. 02: Store the Data

Knowledge Assessment Questions

6) Which of the choices listed below is an option for obfuscation, masking, and anonymization that uses replacement of data with representations of data in its functionality?

A. Masking

B. Randomization

C. Hashing

D. Shuffling

7) With the _____ option, cloud customers using this architecture method are assigned storage in which their data is stored individually as opposed to files.

A. Content Delivery Network

B. Volume Storage

C. Block Storage

D. Object-Based Storage

8) Which of the choices listed below is a responsibility cloud service providers must offer cloud customers associated with the physical plant (datacenter) where the customer data is stored?

A. Use of TPM Standards for BIOS Firmware

B. Virtualization of Management Tools

C. Use of Encryption and Strong Authentication

D. Basing Security Policy on Governance

9) A _____ is a form of data caching typically located near geophysical locations of high consumer demand and data use to increase bandwidth and delivery quality.

A. Content Delivery Network

B. Volume Storage

C. Block Storage

D. Object-Based Storage

10) Which of the choices listed below is the Service Organization Control (SOC) report issued by service providers for their customers as audit assurance statements?

A. SOC 04

B. SOC 03

C. SOC 01

D. SOC 02

True / False Questions

11) The cloud providers' unwillingness to allow customer access to the physical plant facility applies to the customers' auditors and stakeholders as well.

1. True
2. False

12) An area where cloud service providers and customers may find common ground in sharing responsibilities is in the area of compliance auditing and inspections.

1. True
2. False

13) In all cloud service models, the customer and their users will need to access and modify the data at various levels using various permissions and controls.

1. True
2. False

14) The Intrusion Detection System is a large attack surface and offers many potential attack vectors to malicious actors if not secured correctly.

1. True
2. False

15) The cloud provider must apply the proper security controls according to the customers' relevant regulatory industry frameworks and planned usage.

1. True
2. False

16) In addition to securing the hardware components, the cloud service provider must ensure that contractual documents are equally protected for the customer.

1. True
2. False

17) The physical plant of the datacenter will include the facility campus, all physical components, and the services and personnel that support them.

1. True
2. False

True / False Questions

18) Some degree of adversarial relationship exists between the customer and the cloud services provider because they reside in different physical locations.
1. True
2. False

19) An egress monitoring solution will examine data leaving the production environment and react based on preestablished rules and parameters.
1. True
2. False

20) The methods and locations chosen to effectively manage encrypted keys impacts the overall value and benefit of data in several ways.
1. True
2. False

Regulation and Compliance in Cloud Computing

KEY KNOWLEDGE POINTS

Diverse Geographical Legal Jurisdictions
Organizational Cloud Policies
The Cloud in Enterprise Risk Management
Risk Management Frameworks
Risk Management Metrics
Contracts and Service-Level Agreements

Diverse Geographical Locations

A great deal of the difficulty in managing the legal aspects of cloud computing stems from the design of the cloud assets themselves. Cloud assets are dispersed and distributed by necessity, and these assets are being allocated and re-allocated continually making them difficult to identify and control.

The jurisdictions in which the assets are allocated tend to have unique governance, and the legislation within a given jurisdiction is always in flux. Finally, the vagaries of law that are created when technology outpaces the ability of legislatures typically creates slow policy responses.

Organizational Cloud Policies

Policies are a foundational element of organizational governance and risk management programs, and ensure companies operate within their risk profiles. Key stakeholders are usually not directly involved in the creation of organizational policies, but their perceptions can have a dramatic influence on the risk acceptance that leads to the formation of policy.

The creation of policies will change from entity to entity based on organizational structure and a regulatory environment that can limit risk appetite. Listed below are four key stakeholders that typically have the most impact on the creation of cloud computing policies.

1. Board of Directors
2. Senior Leadership
3. Investors and Stakeholders
4. Regulatory and Compliance Entities

Engaging Stakeholders

Identifying and engaging relevant stakeholders is vital to the success of any cloud computing discussions, programs, or projects. Malleable policies should always exist to reflect changes in the organization, but this is especially true when considering the issue of migrating data to the cloud. Cloud risks and benefits are different from their legacy counterparts. For this reason, new policies must be created, and existing policies must be revisited often.

Prioritizing Jurisdictions

The variety and vagaries of multijurisdictional law make the regulatory stakeholders and their input complicated for cloud services. The cloud customers' jurisdiction is earth, making it impossible to craft policy for individual jurisdictions. Additionally, each of these jurisdictions are inherently in conflict (federal vs. state vs. county vs. city vs. the international market). The criteria for prioritizing jurisdictions will change from entity to entity based on organizational structure and geographical location. Listed below are three considerations when choosing a jurisdiction from which organizational policy will be based.

1. Largest Residence of the Providers' End Clientele
2. Location of Most of the Cloud Functionality
3. Jurisdiction with the Most Bearing on Organizational Policy

Communicating Policy

Once a cloud computing policy has been formally accepted, it must be published and disseminated among those affected by the policy. In the best of times, organizational communication is challenging; the issue of cloud computing is increasing that complexity. Providers may be dealing with thousands of customers, and the IT staff of the organization may not be local. Despite this, the internal and external stakeholders should be kept abreast of any changes or issues related to policies and operations.

The methods by which organizations communicate policies will change from entity to entity based on organizational structure and stakeholder needs. Listed below are seven categories of key stakeholders that must always be kept informed.

1. Information Technology and Information Security
2. Human Resource Management
3. Vendor and Client Management
4. Compliance and Regulatory Entities
5. Risk Management and Security Management
6. Finance and Accounting Management
7. Operational Management

Policy Communication Challenges

When discussing these matters with stakeholders, remember they will most likely not have a complete grasp of cloud computing technology. That fact makes it even more crucial for the BCP manager to communicate messages regarding policy clearly and concisely. Good communication will allow senior leadership and key stakeholders to make factual decisions.

Poor communication will create confusion and lead to hearsay. Listed below are five common challenges to communicating policy for BCP managers in organizations with dispersed geographical locations.

1. A Disparate Administrative Workforce
2. Time Zone and Language Differences
3. Ignorance of Cloud Concepts and Models
4. Poor Understanding of Business Drivers
5. Poor Understanding of Risk Appetite

The Cloud in Enterprise Risk Management

It is vitally important that both the cloud customer and the cloud provider focus on risk management and the challenges of cloud computing. The assessment methods and responses to risk will change from entity to entity based on organizational structure and context.

Listed below are the five options an organization has when considering how potential risk will be addressed.

1. Risk Avoidance   (Ignore the Problem)
2. Risk Avoidance   (Reject the Risk Scenario)
3. Risk Acceptance   (Within Risk Appetite Level)
4. Risk Transference   (Insurance and Third Parties)
5. Risk Mitigation   (Never Eliminated Completely)

Risk Management Frameworks: ISO 31000:2009

The *ISO 31000:2009* is an international standard focused on creating, implementing, and reviewing risk management practices and processes. The adoption of this framework will change from entity to entity based on organizational structure and geographical location.

Listed below are seven characteristics of ISO 31000:2009 with which the BCP manager should become familiar before adoption into a risk management program.

1. Integrate Organizational Procedures
2. Engage in Decision-Making Process
3. Explicitly Address Uncertainty
4. Based on Best Available Information
5. Consider Human and Cultural Factors
6. The Program Must be Transparent and Inclusive
7. The Program Must be Dynamic, Iterative, and Responsive

Risk Management Frameworks: NIST SP 800-39

This risk management framework is a methodology for handling all risk in a holistic, comprehensive, and continual manner. It has now superseded the older "Certification and Accreditation" model that was heavily relied upon by federal agencies in the United States.

Although this is an excellent model for domestic operations it is not formally accepted in international markets. This is a crucial fact as some international markets will exclude organizations that are not aligned with ISO standards.

Listed below are four characteristics of NIST SP 800-39 with which the BCP manager should become familiar before adoption into a risk management program.

1. The Framework Relies Heavily on Automation
2. Foundational Purpose is Risk Analysis and Assessment
3. Advocates Assessment-Based Security Controls
4. Advocates Continuous Monitoring and Improvement

Risk Management Frameworks: ENISA

The *European Union Agency for Network and Information Security* is international within Europe, but not globally accepted like ISO standards. It is a comprehensive framework which identifies 35 types of common risks to information systems, and specifically identifies eight security risks based on likelihood.

Listed below are the top five specific risks of the eight identified by ENISA with which the BCP manager should become familiar before adoption into a risk management program.

1. Loss of Program Governance
2. Vendor Lock-In
3. Management Interface Failure
4. Malicious Internal Threats
5. Incomplete and/or Insecure Data Deletion

Risk Management Metrics

To understand whether control mechanisms and policies are effective, it is important to identify metrics that accurately reflect the risk management program.

Levels of risk must be defined and described—quantitatively and qualitatively—and metrics must be attached to the specific risks identified. The levels used to describe risk will change from entity to entity based on organizational structure and context.

Listed below are the five levels of risk that are common to most risk management programs.

1. **Critical**   (Red)
2. **High**   (Orange)
3. **Moderate**   (Yellow)
4. **Low**   (Blue)
5. **Minimal**   (Green)

Contracts and Service-Level Agreements

The most important documents establishing, defining, and enforcing the relationship between the cloud customer and the provider are contracts and *Service Level Agreements* (SLA's).

The terms tend to be used interchangeably but they are distinct and different documents. An important relationship exists between both documents in that they support and corroborate each other.

Listed below are six key characteristics and differences between a contract and an SLA with which the BCP manager should become familiar before accepting the task of reviewing these types of documents.

1. **Contract**: Describes Mutual Responsibilities
2. **Contract**: Defines Services to be Offered
3. **Contract**: Defines Security Provisions to be Provided
4. **Contract**: Stipulates Penalties for Breach of Contract
5. **SLA**: Defines Performance-Based Numerical Metrics
6. **SLA**: Used as a Measurement for Contract Terms

The Service-Level Agreement (SLA)

While both the contract and SLA may contain numerical values, the SLA will expressly include metrics to determine if contractual goals are met. It is important that the cloud customer consider all possible situations and risks associated with cloud business processes and requirements, as this will have a significant impact on negotiating the SLA with the cloud provider.

The metrics included in an SLA to be measured will change from entity to entity based on organizational structure and need.

Listed below are 12 common metrics typically included in service-level agreements for measurement which the BCP manager should become familiar with before accepting the task of reviewing this type of document.

1. Availability Metrics
2. Outage Duration Metrics
3. Capacity Metrics
4. Performance Metrics
5. Storage Device Metrics
6. Server Capacity Metrics
7. Instance Start-Up Time Metrics
8. Response Time Metrics
9. Completion Time Metrics
10. Mean Time to Switchover Metrics
11. Logging and Reporting Metrics
12. Server Scalability Metrics

Regulation and Compliance in Cloud Computing

KNOWLEDGE ASSESSMENT QUESTIONS

The following knowledge assessment questions are presented in true / false, multiple choice, and fill-in-the-blank formats. The correct answers are provided in an Answer Key at the end of Chapter 14. These questions may or may not be presented on quizzes and/or tests given by the instructor of this course.

Knowledge Assessment Questions

1) A great deal of the difficulty in managing the legal aspects of cloud computing stems from the design and distribution of the _____ themselves.

A. Cloud Assets

B. Customer Locations

C. Logistic Centers

D. Governing Agencies

2) Which of the choices listed below is a characteristic of the ISO 31000:2009 risk management framework that distinguishes it from other related risk frameworks?

A. Supersedes the Older "C&A" Model

B. Identifies 35 Types of Risks

C. Not Accepted in International Markets

D. Explicitly Address Uncertainty Factors

3) Policies are a foundational element of governance and risk management programs and ensure companies operate within their chosen _____.

A. Industries

B. Risk Profiles

C. Market Area

D. Geographic Location

4) Which of the choices listed below is a characteristic of the NIST SP 800-39 risk management framework that distinguishes it from other related risk frameworks?

A. The Program is Transparent and Inclusive

B. Considers the Human and Cultural Factors

C. Relies Heavily on Automation

D. 8 Security Risks are Based on Likelihood

5) The variety and vagaries of multijurisdictional _____ make the regulatory entities, stakeholders, and their combined input complicated for cloud services.

A. Policies

B. Customs

C. Laws

D. Clients

Knowledge Assessment Questions

6) Which of the choices listed below is a characteristic of the ENISA risk management framework that distinguishes it from other related risk frameworks?

A. Not Accepted in International Markets

B. Identifies 35 Types of Risks

C. Must be Dynamic, Iterative, and Responsive

D. Relies Heavily on Automation

7) Once an organizational policy has been _____, it must be published and disseminated to those individuals affected by the policy.

A. Edited for Errors

B. Peer Reviewed

C. Initially Drafted

D. Formally Accepted

8) Which of the choices listed below is a characteristic of a Service Level Agreement that distinguishes it from the cloud computing contract in which it resides?

A. Defines Numerical Metrics

B. Stipulates Penalties

C. Describes Responsibilities

D. Outlines Services Offered

9) It is vitally important that both the customer and the cloud provider focus on relevant _____ issues and the challenges of cloud computing.

A. Risk Management

B. Service Agreement

C. Product Marketing

D. Business Development

10) Which of the choices listed below is a factor to be considered while prioritizing jurisdictions when developing organizational cloud computing policies?

A. Identify Jurisdictions with Minimal Impact

B. Crafting Individual Policies is an Option

C. The Smallest Residence of End Clientele

D. The Most Bearing on Organizational Policy

True / False Questions

11) It is important that the cloud provider consider all possible situations and risks associated with cloud business processes, requirements, and potential outcomes.

1. True
2. False

12) While both the contract and SLA may contain numerical values, the SLA will expressly include metrics to determine if contractual goals are being met.

1. True
2. False

13) To understand whether business processes and procedures are effective, it is important to identify metrics that accurately reflect the RM program goals.

1. True
2. False

14) The *European Union Agency for Network and Information Security* (ENISA) is widely accepted within Europe, but not globally accepted like ISO standards.

1. True
2. False

15) The NIST SP 800-39 risk management framework is a methodology for handling all risk in a holistic, comprehensive, and continual manner.

1. True
2. False

16) The ISO 31000:2009 is an international standard focused on designing, implementing, and reviewing vulnerability assessment practices and processes.

1. True
2. False

17) When discussing policy matters with stakeholders, consider that most will not have a complete grasp of cloud computing technology and its implications.

1. True
2. False

True / False Questions

18) Once an organizational policy has been initially conceived and drafted, it must be published and disseminated to those individuals affected by the policy.

1. True
2. False

19) Identifying and engaging relevant stakeholders is vital to the success of any cloud computing discussions, programs, or planned projects.

1. True
2. False

20) Frameworks are a foundational element of governance and risk management programs and ensure companies operate within their chosen risk profiles.

1. True
2. False

CHAPTER 12

Legal Requirements in Distributed Cloud Networks

KEY KNOWLEDGE POINTS

Criminal and Civil Law
Administrative and International Law
Doctrine of Proper Law
EU Data Protection Directive
Personal and Data Privacy Issues
Forensic Requirements

Criminal Law

Criminal law involves all legal matters where the government conflicts with any person, group or entity that violates various statutes. It is enacted by state legislatures and can impose penalties that include fines, imprisonment, or even death. Criminal law applies to state and federal jurisdictions and is adjudicated in various state and federal courts.

States typically handle the prosecution of violations as they have similar criminal laws to the federal system. In those cases in which both federal and state law may be applied in a prosecution, the most stringent laws are the ones typically applied.

Civil Law

Civil law is the body of laws and statutes that deal with personal or community-based law (non-criminal). It is enacted by state legislatures for the purpose of governing private citizens and entities. The implementation of a civil law typically occurs in the form of "lawsuits" or "litigation." Common uses for civil law are breach of contract and adjudication of tort laws (harm resulting from unlawful actions).

Administrative Law

Administrative law is a body of law that affects most people; it is not created by legislatures, but by executive decision and function. Although it impacts the highest majority of people it tends to be the least discussed. Federal agencies representing the Executive Branch of government are the enforcement mechanisms for this category of law.

Listed below are five common issues that administrative law typically adjudicates on a regular basis.

1. Intellectual Property Law
2. Copyrights (Expression of Ideas)
3. Trademarks (Brand Identity)
4. Patents (Inventions and Processes)
5. Trade Secrets

International Law

International law determines how to settle disputes and manage relationships between countries and their respective entities. It is an accumulation of conventions establishing rules agreed upon by all signing parties. International law builds its foundation of practiced customs accepted by law, general principles of recognized law in civilized countries, and judicial decisions establishing precedent of law. The interpretation and application of international law will change from country to country based on internal relationships and political conditions. Listed below are three typical circumstances in which international law is applied.

1. Trade Regulations
2. Tariff Structures
3. Treaties (Solve Disputes and Formalize Alliances)

The Doctrine of the Proper Law

The *Doctrine of the Proper Law* is a term used to describe the processes associated with determining what legal jurisdiction will hear disputes. Additionally, courts will often refer to *Restatement (Second) Conflict of Laws*, a legal concept that keeps courts aware of current legal precedents and decisions relating to jurisdiction.

It is not always a simple matter to determine the jurisdiction in which a case will be heard. For example, consider these geographical complexities regarding a data breach of a major retailer:

1. The Company is Based in the United States
2. Payment Processing is Executed in Canada
3. Customer Goods are Shipped from Singapore
4. Customer Data (PII) is Stored in Ireland

International courts must determine which jurisdiction would hear the case if each entity mentioned above contributed in some way to the breach of customer data. It is a common practice to establish the jurisdiction closest to the area with the largest population of harmed individuals. However, courts may agree to change the venue to a jurisdiction with the most stringent laws.

Relevant U.S. Laws

Stored Communication Act (SCA):

Enacted as part of Title II of the *Electronic Communications Privacy Act of 1986* (ECPA), the SCA addresses both voluntary and compelled disclosure of stored wire and electronic communications and transactional records held by third parties. The ECPA was designed as an extension of the protections previously offered by the *Computer Fraud and Abuse Act* (CFAA) *of 1986*.

Health Insurance Portability and Accountability Act (HIPAA):

The federal *Health Insurance Portability and Accountability Act of 1996* is a set of federal laws governing the handling of personal health information (PHI). The *Office of Civil Rights* (OCR) is the enforcement arm of the *Department of Health and Human Services* (DHHS) and oversees audit findings, policy violations, and unreported breaches.

Gramm-Leach-Bliley Act (GLBA):

The *Gramm-Leach-Bliley Act* (GLBA) also known as the *Financial Services Modernization Act of 1999*, was created to allow banks and financial institutions to merge. It requires institutions to have a written *information security plan* (ISP) and an *information security officer* (ISO) to implement and manage the ISP.

Sarbanes-Oxley Act (SOX):

The *Sarbanes-Oxley Act* was enacted in 2002 as an attempt to prevent unexpected financial collapse due to fraudulent accounting practices, poor audit practices, inadequate financial controls, and poor oversight by governing boards of directors. The *Securities and Exchange Commission* (SEC) is responsible for establishing standards, guidelines, conducting audits, and imposing fines should an aspect of SOX be violated.

EU Data Protection Directive 95/46 EC

The European Union's *Data Protection Directive of 1995*, referred to as the "Data Directive," was the first major EU data privacy law. Unlike the United States—which creates privacy laws and regulations based on industries—the European Union has created privacy laws which are broad, sweeping, and industry agnostic.

Enacted by the European Parliament and Council in 1995, this series of laws was fully implemented in 1998.

In an effort to harmonize privacy laws within member states of the EU, the *General Data Protection Regulation* was enacted in 2016 and made enforceable in 2018.

Although GDPR has officially replaced the "Data Directive," the seven foundational principles of EU privacy law listed below have remained the same.

1. **Notice**: Individual Must be Informed
2. **Choice**: Disclosure of PII is Opt-In, not Opt-Out
3. **Purpose**: Informed of Specific Use of PII
4. **Access**: Individual Allowed Copies of PII
5. **Integrity**: Individual can Correct any Information
6. **Security**: Organizations are Liable for Protecting any PII
7. **Enforcement**: All PII Entities are Subject to EU Authorities

Safe Harbor

The *Privacy Regulation* (GDPR) supersedes the Data Directive and ends the Safe Harbor program, replacing it with a program called "Privacy Shield." The terms of compliance companies handling PII in the EU now face are much more stringent than those originally included in safe harbor. Organizations must agree to auditing and select a federal authority to be responsible for enforcement.

An organization may also agree to GDPR terms by including a formal statement of compliance in their corporate charter under the sections of "Binding Corporate Rules" or "Standard Contractual Clauses." The penalties faced by US-based companies for non-compliance with GDPR tend to be both assured and severe.

Personal and Data Privacy Issues

Due to the decentralized nature of cloud computing, many geographic disparities may present critical personal and data privacy issues. The focus of these concerns will change from entity to entity based on organizational structure and the geographical location of an individuals' PII.

Listed below are six areas of concern BCP managers should take into consideration when assessing the storage of PII in geographically dispersed locations.

1. eDiscovery
2. Chain of Custody
3. Forensic Requirements

4. Direct and Indirect PII Identifiers
5. Contractual and Regulated PII
6. International Conflict Resolution

eDiscovery

Electronic Discovery (eDiscovery) refers to the process of identifying and obtaining evidence for prosecutorial or litigation purposes. Given the centralized nature of cloud computing assets, locating specific records can be challenging.

There are many issues to consider when engaged in eDiscovery operations in multi-tenant environments. Specific customer data must be identified within shared resources—then collected and preserved as evidence—without intruding on third-party data privacy rights.

This can be further impacted by contractual agreements between customers and the cloud service providers.

The Chain of Custody

All evidence needs to be tracked and monitored from the time it is recognized as evidence and acquired for that purpose. BCP managers must incorporate the guidance of professional legal counsel in all chain of custody matters. Strict guidelines apply to the preservation and integrity of evidence.

Only specific and trusted personnel should be involved in the process, and there can be no gaps in the evidence control timeline.

Listed below are four questions regarding evidence collection and chain of custody when performing these operations in cloud computing environments.

1. What People had Access to the Evidence?
2. Where Was the Evidence Stored?
3. What Access Controls were in Place?
4. What Modifications and/or was Analysis Done?

Forensic Requirements

Decentralized data and its movement, storage, and processing across geographic boundaries leads to complex challenges for forensics. Although there is demand for an international forensics standard that would be applicable to all parties and at all locations, no such standard exists currently.

In the absence of such a standard imposed by international laws and treaties, a number of established and accepted standards exist to ease the complexity and burden.

Listed below are four standards approved by the *International Standard for Organization* (ISO) that are providing effective guidance in geographically dispersed locations.

1. ISO/IEC 27037:2012   (Collection)
2. ISO/IEC 27041:2015   (Incidents)
3. ISO/IEC 27042:2015   (Analysis)
4. ISO/IEC 27050:2016**   (eDiscovery)

** ISO/IEC 27050:2016 is currently the most widely accepted standard within the international community.

https://www.iso.org/home.html

Legal Requirements in Distributed Cloud Networks

KNOWLEDGE ASSESSMENT QUESTIONS

The following knowledge assessment questions are presented in true / false, multiple choice, and fill-in-the-blank formats. The correct answers are provided in an Answer Key at the end of Chapter 14. These questions may or may not be presented on quizzes and/or tests given by the instructor of this course.

Knowledge Assessment Questions

1) The _____, also known as the *Financial Services Modernization Act* (FSMA) of 1999, was created to allow banks and financial institutions to merge.

A. Health Insurance Portability & Accountability Act

B. Sarbanes-Oxley Act

C. Stored Communication Act

D. Gramm-Leach-Bliley Act

2) Which of the choices listed below is a distinctive characteristic of administrative law that distinguishes it from criminal, civil, and international law?

A. Created by Executive Decision and Function

B. Enacted by State Legislatures

C. Strictly for Private Entities

D. Addresses Trade Regulations

3) The _____ was enacted in 2002 as an attempt to prevent unexpected financial collapse due to fraudulent accounting practices and poor oversight.

A. Health Insurance Portability & Accountability Act

B. Stored Communication Act

C. Sarbanes-Oxley Act

D. Gramm-Leach-Bliley Act

4) Which of the choices listed below is a distinctive characteristic of criminal law that distinguishes it from administrative, civil, and international law?

A. Created by Executive Decision and Function

B. Created by Federal and State Laws

C. Addresses Breach of Contract

D. Addresses Tariff Structures

5) _____ involves all legal matters where the government conflicts with any person, group or entity that violates various laws and/or statutes.

A. Civil Law

B. Criminal Law

C. Administrative Law

D. International

Knowledge Assessment Questions

6) Which of the choices listed below is a distinctive characteristic of civil law that distinguishes it from administrative, criminal, and international law?

A. Created by Executive Decision and Function

B. Federal Courts Typically Handle the Cases

C. Addresses Tort Laws

D. Addresses Intellectual Property Law

7) The Doctrine of _____ is a term used to describe the processes associated with determining what legal jurisdiction will hear disputes.

A. Proper Law

B. Res Juriscata

C. Transferred Intent

D. Judicial Discretion

8) Which of the choices listed below is a distinctive characteristic of international law that distinguishes it from administrative, criminal, and civil law?

A. Created by Executive Decision and Function

B. Enacted by State Legislatures

C. Addresses Patents and Trademarks

D. Practiced Customs are Accepted as Law

9) The _____ supersedes the *Data Directive* and ends the *Safe Harbor* program, replacing it with a program called *Privacy Shield*.

A. Transparency Regulation

B. Technology Regulation

C. Data Regulation

D. Privacy Regulation

10) Which of the choices listed below is a reason for the *Doctrine of Proper Law* to be considered by legal entities when preparing to hear cases in a court of law?

A. Determine the Legal Jurisdiction

B. Determine the Violated Laws

C. Determine the Affected Parties

D. Determine Fines and Penalties

True / False Questions

11) Decentralized data and its movement, storage, and processing across geographic boundaries creates complex challenges for forensic investigations and standards.

1. True
2. False

12) All contracts need to be secured, tracked, and monitored from the time it is acquired and identified for investigative purposes until the case is closed.

1. True
2. False

13) *Electronic Discovery* (eDiscovery) refers to the process of identifying and obtaining evidence for criminal prosecutorial or civil litigation purposes.

1. True
2. False

14) Due to the centralized nature of cloud computing, many geographic similarities legal parity may present critical personal and data privacy issues.

1. True
2. False

15) The *Privacy Regulation* supersedes the *Data Directive* and ends the *Safe Harbor* program, replacing it with a program called *Privacy Shield*.

1. True
2. False

16) The United States' *Data Protection Directive* of 1995, referred to as the "Data Directive," was the first major national data privacy law.

1. True
2. False

17) The *Doctrine of the Proper Law* is a term used to describe the processes associated with determining what legal jurisdiction will hear disputes.

1. True
2. False

True / False Questions

18) International law is a body of law that affects most people; it is not created by legislatures, but by executive decision and function.

1. True

2. False

19) Criminal law involves all legal matters where the government conflicts with any person, group or entity that violates various laws and/or statutes.

1. True

2. False

20) Administrative law determines how to settle disputes and manage relationships between sovereign, recognized countries and their respective entities.

1. True

2. False

Cloud Computing Contracts and Service Agreements

KEY KNOWLEDGE POINTS

General Contract Structures

The SLA and Contract Review Process

Provisions of an Enforceable Contract

General Contract Review Checklist

Common Myths and Misunderstandings

Contract Negotiation Strategies

** Legal Advice Disclaimer **

The information provided in this text does not, and is not intended to, constitute legal advice; all information is for general informational and educational purposes only. Google is a useful tool for researching legal issues, but Google did *not* pass any states' bar exam. If you have questions regarding advanced topics in technology, always consult a technology expert.

If you have questions regarding legal issues, always consult legal counsel. This content has been created to provide BCP managers with a general understanding of review contracts and service level agreements. However, general understanding is not equivalent to expertise.

Good Fences Make Good Neighbors

The best way to avoid arguments in a business relationship is to write down the parties' expectations ahead of time. Contracts become a boundary marker (like a fence) that explain where the responsibilities of the parties begins and ends. Contracts are not designed for the purpose of winning lawsuits at a later date, nor are they a result of a lack of trust between the parties.

Negotiating contracts can reveal mismatched expectations and sort out details prior to formalizing a relationship. Good contracts actually attempt to prevent disputes and keep the involved parties out of court.

"Legalese" Does Not Exist

Contrary to popular belief, there is no such thing as "legalese," just as there is no "technolese" or "medicalese." Contracts and attorneys may use specialized industry shorthand to convey thought (i.e., *pro forma* or *prima facie*), but so do those in the IT field (*bot*, *dongle*, or *ping*) and it is accepted without a second thought. Even though some attorneys feel the need to use stilted language and long sentences, most modern contracts are written in clear language both parties can understand.

When encountering an unknown legal term, use *Black's Law Dictionary* as a reference or seek out someone with legal expertise. When encountering long sentences that seem to never end, just relax and read slowly.

Seek Best Options Over "Fairness"

Each party negotiating a business relationship has a choice about whether to enter into a contract; neither side owes the other any special considerations or terms. Many individuals believe contracts should be "fair," but it is impossible to define the word in legal terms. Is voluntarily accepting bad terms unfair, and if the majority accepts those terms, are the terms really unfair at all? Focusing on "fair" may cause an individual reviewing a contract to reject terms that make economic sense or accept terms that are ridiculous or absurd. Instead of focusing on "fair," focus on leverage as the guiding principle during contract negotiations and seek the best possible outcome for the relationship.

General Contract Structures

Information Technology contracts can be organized into three groups: *prime clauses, general clauses*, and *boilerplate clauses*. Listed below are the five basic components commonly used in the structure of legal contracts.

1. Introduction and Recitals
2. Definitions
3. Prime Clauses
4. General Clauses
5. Boilerplate Clauses

Prime Clauses

Customers should protect themselves against unclear descriptions and ensure all promises address the expectation of help needed. Listed below are six examples of common issues a BCP manager may expect to review in prime clauses of contracts.

1. Defining the Professional Service
2. Statements of Work (Separate)
3. Change Order Procedures
4. Customer Fees
5. Distributor and Reseller Fees
6. Due Dates and Invoices

General Causes

The term "General Clauses" is a catch-all category that refers to issues not addressed in a primary or boilerplate clause. The one characteristic shared by all elements of general clauses is that they generate the most disagreement, debate, and compromise between the parties involved in the contract negotiation. Listed below are 14 examples of common issues a BCP manager may expect to review in general clauses of contracts.

1. Technical Specifications
2. Service Level Agreements
3. Response, Repair, and Remedy

4. Maintenance, Upgrades, and Updates
5. Schedules and Milestones
6. Delivery, Acceptance, and Rejection
7. Nondisclosure and Confidentiality
8. Data Management and Security
9. General and Specific Indemnity
10. Limitation of Liability
11. Non-Compete and Non-Solicit Stipulations
12. Financial Stability and Reporting
13. Alternative Dispute Resolution
14. Term and Termination of the Contract

Boilerplate Clauses

The boilerplate clauses include a set of terms usually placed at the end of a contract and used for introductory material as well. One cannot predict what issues may arise in a business relationship, so one cannot know when boilerplate clauses will become vital. Listed below are 14 examples of common issues a BCP manager may expect to review in general clauses of contracts.

1. Introductions and Recitals
2. Definitions (Terms Used Often in the Contract)
3. Time Is of the Essence (Time/Breach Clause)
4. Use of Independent Contractors
5. Choice of Law and Courts
6. Government Restricted Rights
7. Technology Export Requirements (International Clients)
8. Force Majeure ("Act of God" and List of Specific Events)
9. Severability (Limits Unknown Impact of Events)
10. Bankruptcy Rights
11. Conflicts Among Attachments
12. Construction (How Unclear Terms are Defined in Courts)
13. Entire Agreement (No Prior Documents)
14. Amendments (Typical and Unilateral)

The SLA and Contract Review Process

Most lawsuits arise from contracts that are not clear, complete, and that do not express the agreement as it was understood by the parties. Listed below are five common guidelines a BCP manager should follow when reviewing contracts and SLA's.

1. Agree on All Definitions
2. Make No Assumptions (Get Clarification)
3. Identify Missing Items (Omissions)
4. Feel Comfortable Changing the Boilerplate as Needed
5. Always Seek a Second Opinion for Reviews

The General Contract Review Checklist

When reviewing contracts and service level agreements, consider the use of checklists to help ensure a consistent approach. The checklists items listed below are not all-inclusive and are only offered as a minimal example of contract review considerations.

1. Ensure all Parties are Properly Identified
2. Ensure all Capitalized Terms are Defined
3. Ensure all Signature and Initial Blocks are Correct
4. Ensure all Referenced Exhibits are Correct and Attached
5. Ensure all Boilerplate is Relevant (Cut and Paste Mistakes)
6. Ensure all Performance Obligations are Accurate
7. Ensure all Payment Terms are Accurate
8. Ensure all Payment Dates and Methods are Accurate
9. Ensure the Service Term and Termination is Correct
10. Ensure the Law, Jurisdiction, and Venue are Correct

Provisions of an Enforceable Contract

A legal contract cannot violate the law in any manner and can become void if either party acts in an illegal fashion. The content of legal contracts will change from entity to entity based on organizational negotiations and legal needs.

Listed below are seven common provisions found in all legal contracts.

1. Capacity (Authorized to Enter Contract)
2. Offer
3. Acceptance
4. Competent Parties
5. Mutuality of Obligation
6. Consideration (Time to Consider Terms)
7. Agreement (Both Parties)

MYTH: Guaranteed Uptime = Guaranteed Uptime

There are many commonly held myths and misunderstandings regarding contracts and service level agreements. One of these misunderstandings can result from contract language defining "guaranteed uptime" of the cloud service. Most providers claim to be operational 99.999% (The Five 9's) of the time, but downtime can mean more than just inaccessible service.

The BCP manager must ensure the true meaning of words, and spot caveats to the conditions which will have a negative impact on the organization. Even when the contract between the customer and provider stipulates subpar performance refunds, those refunds rarely cover the tangible losses experienced by the customer. Listed below are four potential areas of impact a BCP manager should consider when reviewing "guaranteed uptime" performance clauses.

1. **Unreliable or Unusable Service**: The providers' service may be functional, but if it is not functioning at a level useful to the customer it does not mean much. Connectivity to the customer is unreliable or unusable, it might as well be considered downtime.

2. **Service Performance Degradation**: Accessing functions with adequate bandwidth will not provide the customer much value if the function is faulty and there is little or no support to be found to remedy the situation quickly.

3. **Scheduled Maintenance Excluded**: In many contracts, guaranteed uptime does not take into consideration scheduled maintenance (guaranteed downtime). The customer must ensure this issue is addressed contractually.

4. **Iterative Maintenance Plans**: Most major cloud service providers use iterative maintenance plans: incremental maintenance plans rotated across the providers' system. However, this is not guaranteed, and the customer must ensure interactive maintenance plans are in the contract.

MYTH: Contracts Will Scale With Business

This is a common misunderstanding between cloud customers and the providers that can have serious impacts on the organization, but which is easily remedied. The service contract and SLA are designed to meet the needs of the organization at the time they are negotiated and signed, and they do not typically take any future expansion or contraction into consideration.

Organizations can change in size dramatically for a variety of reasons in a short amount of time, and service contracts must take this into consideration. Listed below are five strategies a BCP manager can use to ensure the cloud service contract keeps pace with the needs of the organization.

1. Outline Contract Review Intervals
2. Provider Notification if Customer is Close to Breach
3. Provider Notification if Customers' Service is Underutilized
4. Agreement Negotiations will Begin when Scale Changes
5. Ensure SLA Meetings will Occur "Off Paper" (in Person)

MYTH: Changing Service Providers is Easy

The costs for a customer to change cloud providers can be cost prohibitive and fraught with challenges. A transition clause should be included in the "Term and Termination" section of the cloud contract to avoid overlapping transition costs for the customer, and costly lawsuits for the provider. The transition clause should activate at the time of the contract breach.

For example, if the customer is allocated an "X" amount of data to be transferred each month in a yearly contract, and the provider breaches the contract in month 11 causing the customer to transfer data out to a new provider, will that customer be required to pay "11 x X" for the transfer? Issues such as these must be addressed during the contract negotiations.

MYTH: Providers Should Choose Measurements

It is the organizations' responsibility, not the provider, to determine what performance metrics will be measured in service level agreements tied to cloud service contracts. Most providers are inclined to request measurements in their strong performance areas, while at the same time not revealing their shortcomings to customers. Listed below are five guidelines a BCP manager can use to ensure the proper data is being measured accurately.

1. Decide the Important Metrics for the Company
2. Set Measurable and Specific Benchmarks
3. Avoid Ambiguous Language and/or Numbers
4. Require Monitoring Tools and Timely Reports
5. Demand and Enforce Penalties for Breach of Performance

Contract Negotiation Strategies

Negotiators understand the importance of reaching a win-win solution; when both sides are satisfied, better partnerships are the result. The BCP managers' involvement with contract negotiation will change from entity to entity based on organizational structure and need. Listed below are five proven strategies that should be considered when negotiating cloud service contracts and service level agreements.

1. Make Multiple Offers Simultaneously
2. Include a Matching Right
3. Attempt a Contingency Agreement
4. Negotiate Damages Early in the Process
5. Search for Post-Settlement Settlements

Make Multiple Offers Simultaneously

In negotiations involving many issues, BCP managers can create a great deal of value by making *multiple equivalent simultaneous offers* (MESO). The process begins by identifying several related proposals that are valued equally and presenting the proposals as offers simultaneously. Listed below are three potential benefits of incorporating this action into a negotiation strategy.

1. It Gives Appearance of Flexibility
2. It Increases the Odds of Agreement
3. It may Help to Identify the Oppositions' Preferences

Include a Matching Right

In negotiation strategy, the inclusion of a matching right into a contract offer is a guarantee that one side can match any offer that the other side later receives. This option is gaining popularity in contracts at all levels within a wide variety of industries. For example, if a real estate owner leases office space to a customer for one year and then raises the cost of the rent when the lease expires, the current customer has the right to continue the lease at the higher rate before it can be offered to another tenant.

The grantor (provider) must guarantee this right to the holder (customer) by means of legal contract. This arrangement is also commonly referred to as "Right of First Refusal." Listed below are two potential benefits of incorporating this action into a negotiation strategy for a legal contract.

1. This Right Preserves Mutual Flexibility
2. It Gives the Flexibility to Match Third-Party Offers

Attempt a Contingent Agreement

In negotiation, parties often reach an impasse because they have different beliefs about the likelihood of future events. In this scenario, "Side A" believes a proposal to be reasonable while "Side B" believes the premise upon which the proposal is founded to be unrealistic. If parties involved in contract negotiations cannot align on potential outcomes in the future, there is little chance for any agreement in the present. Listed below are four potential benefits of incorporating this action into a negotiation strategy for a legal contract.

1. Contingent Agreements Are Flexible
2. "If-Then" Challenges are Easier to Overcome
3. Talks can Continue Despite Disagreement
4. Provides an Excellent for Incentives and/or Penalties

Negotiate Damages Early in Process

When formulating the initial contract, the organization can—and should—specify what will happen if any side violates the contract. No matter how well contracts are drafted, lawsuits cause both sides to lose focus on their primary operation. Negotiating damages allows both parties to stipulate the real cost of a breach to the organizations involved (quantitative losses) as opposed to varying perceptions of the cost (qualitative losses). Organizations (and their negotiators) may not feel comfortable leading with this topic but this will benefit both parties.

Listed below are three potential benefits of incorporating this action into a negotiation strategy for a legal contract.

1. Addresses Considerations for Alternate Dispute Resolution
2. Addresses Alternative Means of Compensation
3. Adds a New Issue to the Negotiation (Bargaining Chip)

Search for Post-Settlement Settlements

According to conventional doctrine, the conversation should end once a contract is negotiated for fear of derailing the agreement. That conventional wisdom may not always be practical. Once the contract has been executed there is no harm in testing the other party's willingness to revisit the terms to determine if the existing contract can be improved.

There is no pressure for either side to engage in this type of dialogue. New stipulations and provisions must benefit both parties equally, and either side is free to reject the revised contract if they are so inclined.

Listed below are two potential benefits of incorporating this action into a post-negotiation strategy for a legal contract.

1. It can Identify and Create New Sources of Value
2. It can Lead to Increased Mutual Trust Between Parties

Cloud Computing Contracts and Service Agreements

KNOWLEDGE ASSESSMENT QUESTIONS

The following knowledge assessment questions are presented in true / false, multiple choice, and fill-in-the-blank formats. The correct answers are provided in an Answer Key at the end of Chapter 14. These questions may or may not be presented on quizzes and/or tests given by the instructor of this course.

Knowledge Assessment Questions

1) The best way to avoid arguments in a business relationship is to _____ the parties' expectations ahead of time, then understand and agree to them.

A. Write Down

B. Negotiate

C. Reject

D. Demand

2) Which of the choices listed below is a distinctive characteristic of Prime Clauses that distinguishes them from other sections in general legal contract structures?

A. Includes Service Level Agreements

B. Includes Introductions and Recitals

C. Addresses Nondisclosure and Confidentiality

D. Used for Defining the Professional Service

3) Cybersecurity and Information Technology contracts can be organized into three groups: *prime clauses,* _____, and *boilerplate clauses.*

A. Statements of Work

B. General Clauses

C. Service Level Agreements

D. Amendments

4) Which of the choices listed below is a component of an enforceable contract which must be present for the document to be considered legal and acceptable?

A. Filing Time Stamp

B. Letterhead

C. Mutuality of Obligation

D. Notary Seal

5) You cannot predict what issues may arise in a business relationship, so you cannot know when the content of _____ clauses will become vital.

A. Severability

B. General Clauses

C. Boilerplate

D. Prime

Knowledge Assessment Questions

6) Which of the choices listed below is a distinctive characteristic of General Clauses that distinguishes them from other sections in general legal contract structures?

A. Includes Statements of Work

B. Includes Service Level Agreements

C. Identifies Choice of Law and Courts

D. Includes Amendments

7) In negotiation, include a _____ in your contract — a guarantee that one side can match any offer that the other side later receives.

A. Post-Settlement

B. Contingent Agreement

C. Disclaimer

D. Matching Right

8) Which of the choices listed below is a reason to seek post-settlement settlements when negotiating the details of legal contracts and formal agreements?

A. Can Lead to Increased Mutual Trust

B. Considers Alternative Dispute Resolution

C. Talks can Continue Despite Disagreement

D. Addresses "Right of First Refusal"

9) When reviewing contracts and service level agreements, consider the use of _____ to help ensure a consistent approach for evaluation.

A. Standardized Checklists

B. Peer Review

C. Second Opinions

D. Legal Dictionaries

10) Which of the choices below is a distinctive characteristic of Boilerplate Clauses that distinguishes them from other sections in general legal contract structures?

A. Includes Change Order Procedures

B. Includes Schedules and Milestones

C. Addresses Limitation of Liability

D. Includes Introductions and Recitals

True / False Questions

11) Customers should protect themselves against unfair contracts and ensure all promises and commitments address the expectation of service needed.
1. True
2. False

12) The needs of the organization do not always align with the desires of the service provider. For that reason, measurements should be chosen by the client.
1. True
2. False

13) The true costs associated with changing service providers may not always be immediately clear. For that reason, advance payment is a requirement.
1. True
2. False

14) Words are important, and clear definitions even more so. Ensure what is being stated in the contract accurately reflects the reality of the situation.
1. True
2. False

15) Most disagreements arise from contracts that are not clear, complete, and that do not express the agreement as it was understood by the courts involved.
1. True
2. False

16) The *Boilerplate Clauses* include a set of terms usually placed at the end of a contract and are also used for introductory material as well.
1. True
2. False

17) The one characteristic shared by all elements of *General Clauses* is that they generate the most agreement, consensus, and generous terms.
1. True
2. False

True / False Questions

18) When formulating the initial contract, you can — and should — specify what will happen if either side violates the terms of the contract.

1. True

2. False

19) In negotiation, parties often reach an impasse because they have similar beliefs about the likelihood of future events being discussed at the time.

1. True

2. False

20) Each party has a choice about whether to enter and agree to a contract; neither side owes the other any special considerations or terms during the process.

1. True

2. False

Testing and Improving Disaster Preparedness Plans

KEY KNOWLEDGE POINTS

Developing Testing Goals
Benefits of BCP Testing
BCP Test Progressions
Potential Testing Scenarios
Training and Awareness
Benefits of Certifying the BCP

Writing a Testing Strategy

To maximize the benefit to the organization while minimizing costs, develop a written testing strategy for the BCP. The testing strategy will be a written document in the Administrative BCP describing the types and frequency of testing. The testing strategy will have formal approval of senior leadership (as do all documents in the Administrative BCP) which serves to establish a higher level of compliance with the plan.

Calendars are typically planned for a span of several years. The BCP manager must consider that organizational units have a variety of "busy seasons" that would make participation difficult. Initiate the testing strategy with an individual plan, then expand it over time to include multiple BCP groupings.

Developing Testing Goals

As with all initiatives related to the BCP, begin setting goals for the testing plan by referring to the BIA. Goals will change from entity to entity based on organizational structure and needs. Listed below are six points relating to testing goals of threats typically considered by BCP managers.

1. **Brief RTO** = Testing is More Frequent
2. **Long RTO** = Testing is Less Frequent
3. Incident Severity can Change Existing RTO's
4. Specific Process RTO's can Differ
5. Critical Processes are Tested Frequently
6. All Participants must Understand Their Roles

Creating the Testing Team

The best BCP testing results come from a clear explanation of the responsibilities of team members, and training to show them what to do. Listed below are five categories of participants that typically make up a BCP testing team.

1. The Business Continuity Manager
2. The Test Sponsor
3. The Exercise Recorder
4. The Exercise Participants
5. External Non-Organizational Participants

Leading BCP Testing Exercises

Leading a testing exercise takes planning, preparation, and a complete understanding of expected outcomes and goals. Choosing a relevant test scenario can be difficult contingent upon the specific needs of the participants. The scenario should be focused on the type of problem that a specific group of people are likely to face.

Test scenarios will change from entity to entity based on organizational structure and needs. Listed below are seven points typically considered by BCP managers who intend to lead testing exercises.

1. Scenarios Reflect Real and Potential Threats
2. Use Risk Analysis and Risk Assumptions in Planning
3. The Expertise of the Coordinator is Critical
4. Create a Detailed Timeline of Events
5. Allocate, Coordinate and Stage all Resources
6. Establish and Assign Specific Participant Roles
7. Identify and Address Relevant Interdependencies

The Benefits of BCP Testing

An untested plan is nothing more than process documentation. During times of crisis, people tend to react the way they were trained, and perception tends to become the reality. A test that is conducted well and thoroughly can have a positive influence on the participants. Listed below are five benefits that typically result from well-conducted BCP testing exercises.

1. Reveal Plan Errors and/or Incorrect Assumptions
2. Uncover Changes Made Since the Plan was Written
3. Reveal Missing and/or Unnecessary Steps
4. Identify Unknown Contingencies
5. Verify Resource and/or Asset Availability

The 5 Progressions of BCP Testing

The 5 levels of BCP testing allow for incremental improvements over time and a wider variety of testing to be conducted more often. Listed below are five progressions of BCP tests typically conducted by organizations. These tests will be discussed in further detail in the following pages.

1. Checklist Test
2. Table-Top Test
3. Structured Walk-Through
4. Parallel Test
5. Fail-Over Test (Full Interruption)

Progression 01: Checklist Testing

Checklist testing is the first (and easiest) level of the BCP testing progression. The objective of this test is to create and work through a BCP that is standardized and understood by all the participants. The administration of this test will change from entity to entity based on testing goals and organizational needs.

Listed below are five characteristics typical of this type of test.

1. Required After Significant System Changes are Made
2. Tests Individual IT and Business Processes
3. The First Level of Error-Checking for Plans
4. The Objective is Complete Recovery from Nothing
5. Ensures that the Recovery Plan is Understandable

Progression 02: Table-Top Testing

Table-top testing is the second level of BCP testing progression. The objective of this test is to train specific team members, identify omissions in the plan, and raise general awareness of the BCP.

Listed below are five characteristics typical of this type of test.

1. Initiated as a Simulated Emergency Scenario
2. A Test of the Participants' Decision-Making Processes
3. Focuses on Analysis, Communication, and Collaboration
4. Includes Mid-Exercise Problem Injections
5. Typically Conducted as a Half-Day Exercise

Progression 03: The Structured Walk-Through

The *structured walk-through* is the third level of the BCP testing progression. The objective of this test is to focus on specific IT assets and is typically conducted in the environment in which the assets are located.

Listed below are five characteristics typical of this type of test.

1. Tests Multiple Plans in Logical Groupings
2. Ensures Effective Data Exchange and Communication
3. Test the Strength of Asset Interdependencies
4. Can Utilize Disaster Recovery Site Machines and Systems
5. Beneficial in Determining Realistic RTO's

Progression 04: Parallel Testing

Parallel testing is the fourth level of the BCP testing progression and tends to be more complex than the first three progressions. The objective of this test is to evaluate the BCP with actual systems at the recovery site without the risk of interrupting the operation of the organization.

Listed below are five characteristics typical of this type of test.

1. Conducted as a Realistic Emergency Scenario
2. Tests Multiple and Logical Groups of Plans Simultaneously
3. Includes Key Personnel and Recovery Site Systems
4. Actions are Conducted at the Disaster Recovery Site
5. The Exercise Typically Runs for Several Days

Progression 05: Fail-Over Testing

Fail-over testing (full interruption) is the fifth level of the BCP testing progression and is by far the riskiest of all the BCP test progressions. The objective is to create a full interruption of the organizations' system to definitively test the BCP.

Listed below are five characteristics typical of this type of test.

1. There is a Full Shutdown of the Primary Location
2. There is a Full Activation of the Secondary Site
3. This Test Creates 100% Confidence in the BCP
4. A Suitable Test for Mature and Redundant Systems
5. The Exercise Typically Runs for Several Days

Potential Testing Scenarios

Realistic testing scenarios based on recent events tend to generate more "buy in" from BCP testing participants. The testing coordinator must also ensure any scenario selected will have relevant and achievable outcomes. The potential testing scenarios will change from entity to entity based on testing goals and the organizations' location and needs. Listed below are four examples of testing scenarios that replicate real-world events.

1. Natural Disasters
2. Civil Crisis and Unrest
3. Geographic Location and Proximity Threats
4. Network and Information Security Threats

Improvisational Exercises

Planned events can create unique opportunities to execute testing scenarios that are relevant to the business operation. It is the BCP managers' responsibility to be aware of these scheduled events to ensure enough time is available to allocate necessary resources and plan the exercise.

Listed below are five examples of improvisational testing opportunities that typically exist for most organizations.

1. Scheduled Power Outages (Maintenance or Conservation)
2. Facility Construction Projects
3. Business Unit and/or System Relocation
4. New Facility Acquisition (Pre-Occupancy)
5. Local Government Disaster Drills (Published and Open)

Demonstrating RTO Capability

The RTO for any function or process is typically based on a perceived need identified by an organization. However, that need is not always consistent with the ability to achieve the RTO. BCP tests allow for actual RTO's to be documented and measured under the best and worst of circumstances.

BIA estimates are never completely accurate despite best efforts; BCP testing measures written theory against documented reality. It also allows the BCP manager to consider parallel tasking options, reevaluate tasking sequences, and identify (then eliminate) non-essential tasks.

This type of analysis will determine if the RTO can be achieved with a more streamlined approach, or if it must be adjusted to reflect more realistic expectations and outcomes.

Debriefing Exercise Participants

Once any progression of BCP testing has been concluded, the next critical step for the BCP manager is to debrief all participants involved with any aspect of the testing. This is a two-way exchange of information between the test facilitators and participants. *This type of debriefing is not restricted to BCP testing alone.*

Whenever an incident occurs in the organization that is covered by the BCP, the BCP manager should conduct an after-action review on the next workday after the recovery. The debriefing style and timeframes will change from entity to entity based on the organizations' policy and needs.

Listed below are five topics typically included in any BCP test debriefing.

1. **What Happened?** Different participants will have different perceptions contingent upon what role they were assigned. Collecting this information will provide the BCP manager with a holistic view of the testing event in its entirety.

2. **What Should Have Happened?** Each participant assigned a role in the testing scenario knew the expectations going into the exercise, observed what occurred during the test, and will be able to identify any gaps that existed. Collecting this information will provide the BCP manager with an accurate idea of what areas of the plan need to be revisited for improvement.

3. **What Went Well?** Only on the rarest occasions will every objective of a test be missed. Collecting this information will provide the BCP manager with an accurate assessment of the plan items that worked, even if they can be better.

4. **What Did Not Go Well?** This will be the easiest question to answer (in theory) but the most difficult for the test participants to articulate (in reality). Hearing opinions is important, but the BCP manager should stay focused on objective analysis and ensure participants do not engage in finger-pointing or the "blame game."

5. **What Will We Do Differently?** Although this question is typically answered after all information has been collected, there are some actionable items which will be obvious to all participants immediately after the test ends. This is a suitable time for the BCP manager to solicit feedback and positive suggestions from everyone involved with the test.

Participant Testing Considerations

When BCP testing outcomes fall short of expectations, especially during high-intensity scenarios, emotions of the participants can increase exponentially both for better and for worse. The BCP manager must be aware of this during the initial debriefing and the days following the test as well.

Ignoring the emotions and concerns of test participants will discourage participation in future tests. Listed below are seven suggested guidelines for BCP managers to follow when considering the emotional state of test participants.

1. Praise in Public
2. Criticize in Private
3. Do not Assign Blame to Participants
4. Discovering Gaps is not Equal to Failure
5. Keep Observations and Analysis Objective

6. Remove Judgment from any Consideration
7. Respect, Acknowledge, and Document Dissenting Opinions

BCP Training and Awareness

Training and awareness are the core of successful BCP testing programs; people will typically act (and react) in the manner they were trained. The training methodologies and mediums will change from entity to entity based on the organizations' policy and needs.

Listed below are six typical characteristics incorporated into successful BCP training programs.

1. Identify Clear Training Goals and Outcomes
2. Make the Training Engaging, Fun, and Competitive
3. Focus on Changing Behaviors and Perceptions
4. Strive to Bridge the "Knowledge vs. Action" Gap
5. Tailor Appropriate Training to the Target Audience
6. Solicit Participant Ideas and Encourage Feedback

The Benefits of Certifying the BCP

Third-party certification of BCP plans is not a requirement, but it does have benefits that may justify this expense to the organization. BCP testing can be unintentionally structured to focus on a plans' strengths while glossing over weaknesses; professional certification addresses this issue using objective and qualified third parties.

Examples of certification organizations include ISACA and the International Organization for Standardization and are typically selected based on the organizations' industry and compliance needs. Listed below are five potential benefits that third-party certifications may provide to a BCP program.

1. Objective External Reviews can Reveal Weaknesses
2. Process Ensures Alignment with Best Practices
3. Certification Provides a Higher Degree of Confidence
4. Certification Provides a Higher Degree of Credibility
5. Certification can Fulfill a Responsibility to Stakeholders

Certification Resources

The organization chosen to certify the Business Continuity Plan will depend on the type of industry for which the plan is being written. There is no organization considered to be the best; each industry has their own preferred certifying body.

Listed below are links to two well-established and well-established organizations that are frequently used to evaluate and validate BCP / DRP programs.

ISACA CMMI Performance Solutions

https://cmmiinstitute.com

International Organization for Standardization (ISO)

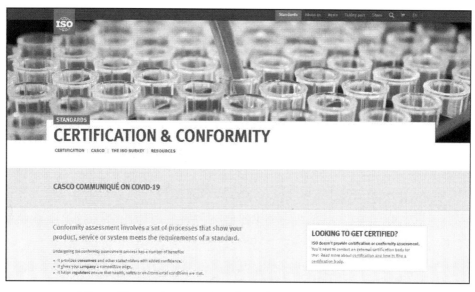

https://www.iso.org/conformity-assessment.html

Testing and Improving Disaster Preparedness Plans

KNOWLEDGE ASSESSMENT QUESTIONS

The following knowledge assessment questions are presented in true / false, multiple choice, and fill-in-the-blank formats. The correct answers are provided in an Answer Key at the end of Chapter 14. These questions may or may not be presented on quizzes and/or tests given by the instructor of this course.

Knowledge Assessment Questions

1) As with all things in the Business Continuity program, begin setting goals for the testing plan by referring to the _____.

A. Business Impact Analysis

B. Risk Management Framework

C. Corporate Security Policy

D. Local Emergency Management

2) Which of the choices listed below is a distinctive characteristic of *Checklist Testing* that distinguishes it from other progressions of BCP / DRP testing scenarios?

A. Tests the Logical Grouping of Plans

B. Utilized to Determine a Realistic RTO

C. Incorporates Mid-Exercise Problem Injections

D. The First Level of Plan Error Checking

3) Testing can be structured to focus on a plans' strengths and gloss over weaknesses. Third-party _____ seeks to overcome this issue.

A. Involvement

B. Certification

C. Outsourcing

D. Opinions

4) Which of the choices listed below is an example of a planned external event which can be used to facilitate improvisational BCP / DRP testing opportunities?

A. Natural Disasters

B. Civil Unrest

C. Local Government Disaster Drills

D. Organizational Emergencies

5) Use blogs, newsletters, posters, and email tips to maximize the messaging of training. _____ training tends to be much more engaging.

A. Well-Attended

B. Mandatory

C. Highly-Interactive

D. Quarterly

Knowledge Assessment Questions

6) Which of the choices listed below is a distinctive characteristic of *Parallel Testing* that distinguishes it from other progressions of BCP / DRP testing scenarios?

A. Used to Create 100% Confidence in the Plan

B. Includes Relevant Personnel and Equipment

C. Ensures the Recovery Plan is Understandable

D. The First Level of Plan Error Checking

7) Realistic testing scenarios based on _____ tend to generate more "buy in" from BCP testing participants and encourage active engagement.

A. Leadership Suggestions

B. Popular Shows

C. Employee Experience

D. Recent Events

8) Which of the choices listed below is an example of a BCP / DRP team member who would be expected to attend any organizational BCP testing exercise?

A. Business Continuity Manager

B. Court Recorder

C. Corporate Sponsors

D. Unsolicited Volunteers

9) Whenever an incident occurs that is covered by the BCP, conduct a(n) _____ on the next workday after the recovery with participants involved.

A. After-Action Review

B. Press Conference

C. Management Presentation

D. Training Seminar

10) Which of the choices listed below is a distinctive characteristic of *Fail-Over Testing* that distinguishes it from other progressions of BCP / DRP testing scenarios?

A. The First Level of Plan Error Checking

B. May Utilize Recovery Site Systems

C. Ensures the Recovery Plan is Understandable

D. A Full Shutdown of the Primary Location

True / False Questions

11) To maximize benefits to the organization while minimizing costs, develop a written testing strategy for the BCP that can be planned and revised in advance.

1. True
2. False

12) The best BCP testing results come from a clear explanation of the responsibilities of unit managers and providing training to demonstrate the plan expectations.

1. True
2. False

13) The scenarios selected should be focused on the types of problems that a specific group of people within the organization are likely to encounter.

1. True
2. False

14) The 6 progressions of BCP testing allow for exponential improvements over time and a smaller variety of testing to be conducted less often.

1. True
2. False

15) When testing exercise outcomes fall short of expectations, do not blame participants. Blame and fault are not productive and discourage participation.

1. True
2. False

16) Unplanned internal events can create unique opportunities to execute testing scenarios that are relevant to various aspects of the business operation.

1. True
2. False

17) The objective of a *Fail-Over Test* is to create a full interruption of organizational systems to definitively test the BCP implementation and the personnel involved.

1. True
2. False

True / False Questions

18) The objective of a *Table-Top Test* is to create and work through a BCP that is standardized and understood by all participants involved in the process.

1. True
2. False

19) A *Structured Walk-Through* is conducted in the actual test environment and typically focuses on specific IT systems, personnel, and potential challenges.

1. True
2. False

20) The objective of a *Parallel Test* is to create and work through a BCP that is standardized and understood by all participants involved in the process.

1. True
2. False

Knowledge Assessment
ANSWER KEY

Knowledge Assessment Answer Key

1) "*A set of policies, tools, and procedures to enable the recovery and continuation of mission-critical technology infrastructure and systems following a natural or human-induced disaster*" is the definition of _____.

A. Business Continuity

B. Incident Response

C. Risk Mitigation

D. Disaster Recovery

2) Which choice below is a goal of information security professionals when mapping the Business Continuity Plan to organization objectives?

A. Increase Employee Benefits

B. Maintain a Corporate Profit Margin

C. Decrease Competitive Advantage

D. Avoid Regulatory Compliance

3) Business Continuity and Disaster Recovery plans must account for compliance, _____, and operational process alignment to be successful.

A. Risk Management

B. Vulnerability Assessment

C. Financial Services

D. Organizational Leadership

4) What statement listed below represents an accurate characteristic of an Epidemic / Pandemic Continuity Plan?

A. Affects People at Some Levels

B. Only Considers Internal Dependencies

C. Different from Other Recovery Plans

D. Affects the Organization at Some Levels

5) "*The maximum acceptable amount of data loss, measured in time, that can be incurred without serious and/or negative impact to the organization*" is the definition of _____.

A. Recovery Time Objective

B. Recovery Point Objective

C. Maximum Allowable Downtime

D. Quantitative Analysis

Knowledge Assessment Answer Key

6) What statement listed below represents an accurate characteristic of an Incident Response Plan from the perspective of Business Continuity?

A. Details Remediation and Eradication Steps

B. Addresses the Identification of Events

C. Establishes Financial Priorities and Actions

D. Defines the Secure Development Life Cycle

7) The _____ is a consolidated, high-level document written to meet a set of specific business needs.

A. Administrative Continuity Plan

B. Technical Continuity Plan

C. Crisis Management Plan

D. Incident Response Plan

8) What choice below can be described as a documented component of a Vulnerability Assessment Plan within a Business Continuity strategy?

A. Incident Escalation Procedures

B. Malware Triage and Analysis

C. Collection and Preservation of Evidence

D. Threat Intelligence Collection and Analysis

9) The _____ is a detailed document, maintained and managed by BCP leadership, written to address a set of critical workflow needs.

A. Administrative Continuity Plan

B. Digital Forensics Plan

C. Work Area Continuity Plan

D. Technical Continuity Plan

10) What choice below describes a real-world challenge faced by organizations when activating Disaster Response and Recovery Plans?

A. The Level of Impact is Known

B. Initial Information is Typically Minimal

C. Access to Sites can be Unrestricted

D. Initial News Reports can be Informative

True / False Answer Key

11) Information Security is not a business objective in and of itself, but Information Security underlies all business objectives in every organization.

True

12) The *Business Continuity Plan* (BCP) is a monolithic and linear document. It is supported bottom-up, created top-down, and changes occasionally.

False

The *Business Continuity Plan* (BCP) is not a monolithic or linear document. It is supported top-down, created bottom-up, and always evolving.

13) The *Disaster Recovery Plan* (DRP) is not a monolithic or linear document. It has many moving parts that are executed simultaneously.

True

14) The first draft of a policy is usually adequate, and the acceptance level of policies tends to be widespread by nature and evolve with financing over time.

False

The first draft of a policy is rarely adequate, and the maturity level of policies tends to be incremental by nature and evolve with experience over time.

15) There are many common obstacles when defining a desired security state which must be acknowledged and defined in both business and security terms.

True

16) The *Digital Forensics Plan* is a detailed document written to meet a set of specific response and mitigation needs during an adverse situation.

False

The *Crisis Management Plan* is a detailed document written to meet a set of specific response and mitigation needs during an adverse situation.

17) The *Technical Continuity Plan* is detailed a document, maintained and managed by IT leadership, written to address specific technology needs.

True

True / False Answer Key

18) Design solutions that account for real-world challenges, not unrealistic solutions which only serve to appease senior management or look good on paper.
True

19) An incident is any user- or system-generated action or occurrence that can be identified by a program and has significance for system hardware or software.
False

An event is any user- or system-generated action or occurrence that can be identified by a program and has significance for system hardware or software.

20) Qualitative analysis is the examination and evaluation of non-measurable data using subjective judgement and non-quantifiable methods.
True

Knowledge Assessment Answer Key

1) A _____ provides many benefits to the organization which are valuable beyond the scope of a Business Continuity Planning project.

A. Vulnerability Assessment

B. Financial Projection Analysis

C. Risk Assessment

D. Business Impact Analysis

2) What choice below represents a best practice to be followed when conducting formal presentations at all levels of the organization?

A. Limit Bullet Points and Text

B. Use Creative Fonts

C. Use Popular Templates

D. Use Animated Graphics

3) An effective _____ process will help quantify the value of each function in terms of financial and legal impacts.

A. Qualitative Analysis

B. Asset Distribution

C. Data Collection

D. Quantitative Analysis

4) What choice below represents a potential consequence of a negative impact to an organizations' "goodwill"?

A. Maintaining Competitive Advantage

B. Damage to Brand or Image

C. Increase in Product Revenue

D. Restricted Hiring Budgets

5) Once the Business Impact Analysis _____ have been developed and tested in a single unit, distribute them to all appropriate business units.

A. Reports

B. Questionnaires

C. Budgets

D. Approvals

Knowledge Assessment Answer Key

6) What choice below represents a best practice to be followed when initially developing a Business Impact Analysis questionnaire?

A. Test Questionnaires in Multiple Units

B. Standardize Questions for Business Units

C. Solicit Feedback and Clarifications

D. Limit Options to Answer Questions

7) Check _____ and travel schedules for all selected respondents to help ensure timely completion of the Business Impact Analysis questionnaires.

A. Vacation

B. Training

C. Meeting

D. Presentation

8) What choice below can be described as a benefit of conducting a Business Impact Analysis in support of a Business Continuity Plan?

A. Identify Unprofitable Functions to Cancel

B. Confirm Recovery Time Objectives

C. Identify and Prioritize Abundant Resources

D. Identify Vital Business Records

9) Create a formal process for exemptions and _____ that documents decisions and validates the acceptance of risk by senior leadership.

A. Key Employees

B. Risk Ratings

C. Questionnaires

D. Exceptions

10) What choice below represents a potential tangible cost to an organization which must be considered when conducting a Business Impact Analysis?

A. Legal Penalties for Non-Compliance

B. Loss of Customer Goodwill

C. Reduced Shareholder Confidence

D. Damage to the Brand Image

True / False Answer Key

11) To ensure the highest probability of a successful outcome, preparation and simplicity are the best methods to present compiled results to decision-makers.

True

12) Consider creating a temporary staffing list to anticipate and prepare for key personnel considerations which may emerge at the last minute unexpectedly.

False

Consider creating an "Employee Skills Matrix" to anticipate and prepare for key personnel considerations which may emerge at the last minute unexpectedly.

13) The first step in identifying who should receive the BIA questionnaire is to secure a current copy of the organizational chart.

True

14) Create an informal process for exemptions and exceptions that documents decisions and validates the rejection of risk by senior leadership.

False

Create a formal process for exemptions and exceptions that documents decisions and validates the acceptance of risk by senior leadership.

15) Once the BIA questionnaires have been developed and tested in a single unit, distribute them to all appropriate business units.

True

16) An effective BIA data collection process will help qualify the value of each function in terms of staffing and productivity impacts.

False

An effective BIA data collection process will help quantify the value of each function in terms of financial and legal impacts.

17) The Business Impact Analysis project must be supported financially and politically from the highest levels of the organization.

True

True / False Answer Key

18) Tangible costs due to the loss of a vital business function can be more difficult to identify but are less damaging than intangible costs.
False

Intangible costs due to the loss of a vital business function can be more difficult to identify but are no less damaging than tangible costs.

19) There are numerous ways the loss of a critical function can have a negative financial impact on the organization.
True

20) Ensure the right people are asked the right questions regarding goodwill. There are many intangible costs with tangible consequences.
True

Knowledge Assessment Answer Key

1) *"A probability or threat of damage, injury, liability, loss, or any other negative occurrence that is caused by external or internal vulnerabilities, and that may be mitigated through preemptive action"* is the definition of _____.

A. Risk

B. Vulnerability

C. Likelihood

D. Severity

2) Which of the choices listed below is a component and best practice of the risk management process?

A. Understand Previous Threats

B. Determine Potential Vulnerabilities

C. Accept the Levels of Risk

D. Review Control Effectiveness

3) *"The evaluation and estimation of the levels of risks involved in a situation, their comparison against benchmarks or standards, and the determination of the probable severity of their impact"* is the definition of _____.

A. Risk Appetite

B. Risk Assessment

C. Vulnerability Assessment

D. Residual Risk

4) Which of the choices listed below is a characteristic to consider when creating an asset resource profile?

A. System Type and Version

B. Logical Network Addressing

C. Functions and Features

D. Vendor Support Information

5) Although risk management program development consists of 6 high-level steps, do not lose focus or awareness on the eventual details and procedures of _____.

A. Approval

B. Financing

C. Implementation

D. Acceptance

Knowledge Assessment Answer Key

6) Which of the choices listed below is a characteristic of the OCTAVE Allegro risk analysis framework and methodology?

A. Requires Pre-Planning by a Facilitator

B. Uses a Three-Phased Approach

C. Based on Two Factors to Measure Risk

D. Establishes Accountability for Controls

7) A(n) _____ provides critical information for the technical BCP and is typically kept in both the BCP and the asset location.

A. Resource Profile

B. Personnel Log

C. Gantt Chart

D. Asset Inventory

8) Which of the choices listed below is a characteristic of the Facilitated Risk Analysis and Assessment Process (FRAAP)?

A. Focused on Project Level Assessments

B. Real-Time Continuous Monitoring

C. Creates a 6-Level Risk Exposure Table

D. Uses Highly Detailed Questionnaires

9) The identification and costs of _____ risks is highly beneficial to a thorough risk analysis. Be aware these risks exist even if they cannot be controlled.

A. Manufactured

B. Transferred

C. Mitigated

D. Residual

10) Which of the choices listed below is a characteristic of the Factor Analysis of Information Risk (FAIR) framework and methodology?

A. Building Asset-Based Threat Profiles

B. Automated Tools to Inform Decisions

C. A Streamlined Approach Using SME's

D. Probable Loss Magnitude (PLM)

True / False Answer Key

11) Security managers should be involved in all phases of the SLA process and ensure supporting providers have adequate controls.

True

12) Risk management frameworks and analysis methodologies provide value to selected processes if their primary focus is technical and data driven.

False

Risk management frameworks and analysis methodologies provide value to the overall process whether their primary focus is technical or non-technical.

13) When choosing a risk management framework, adapt and utilize established reference models based on organizational culture and specific needs.

True

14) Risk assessment considers the basic factors of asset value, both tangible and intangible, within several contexts and potential scenarios.

False

Asset valuation considers the basic factors of asset value, both tangible and intangible, within several contexts and potential scenarios.

15) An asset resource profile is supporting confidential information for risk mitigation. Document the resource profile with the asset inventory information.

True

16) There are numerous frameworks and industry tools available for assessing risk, but the components of vulnerabilities and threats share many similarities.

False

There are numerous frameworks and industry tools available for assessing risk, but the components of the risk management process share many similarities.

17) Risk appetite is the level of risk that an organization is prepared to accept in pursuit of its objectives, and before action is necessary to reduce the risk.

True

True / False Answer Key

18) Risk analysis is a relative measurement of a resources' tolerance for risk exposures, independent of any threat or vulnerability.

False

Risk sensitivity is a relative measurement of a resources' tolerance for risk exposures, independent of any threat or vulnerability.

19) Not all third-party service providers can be audited, but for those that can the audit requirement should be negotiated into the Service Level Agreement (SLA).

True

20) Risk assessment is a useful tool to identify weaknesses that may be visible to threat sources.

False

Vulnerability assessment is a useful tool to identify weaknesses that may be visible to threat sources.

Knowledge Assessment Answer Key

1) Most _____ analysis approaches use a relative scale to rate risk exposures based on a set of predefined criteria established for each level.

A. Qualitative

B. Predictive

C. Statistical

D. Quantitative

2) Which of the choices listed below is a component used in the calculation of Single Loss Expectancy (SLE)?

A. Exposure Factor

B. External Impact

C. Annualized Rate of Occurrence

D. Annualized Loss Expectancy

3) It is not uncommon to confuse factors that will affect the severity and likelihood of the threat / _____ pair.

A. Response

B. Exploitation

C. Vulnerability

D. Risk

4) *"An analysis method which uses subjective judgment to analyze a company's value or prospects based on non-quantifiable information"* is the definition of

_____.

A. Risk Analysis

B. Qualitative Analysis

C. Threat Analysis

D. Quantitative Analysis

5) The severity rating, commonly used in qualitative analysis models, is rated by the degree of potential _____.

A. Risk

B. Disruption

C. Likelihood

D. Consequences

Knowledge Assessment Answer Key

6) Which of the choices listed below is a factor to be considered when rating the severity of data availability?

A. Deletion of Data

B. Modification of Data

C. Degradation of Performance

D. Unauthorized Access

7) When defining risk scales for data availability severity it is important to ensure there are visible distinctions between _____.

A. Levels

B. Assets

C. Threats

D. Vulnerabilities

8) *"An analysis method of a situation or event, especially a financial market, by means of complex mathematical and statistical modeling"* is the definition of _____.

A. Risk Analysis

B. Qualitative Analysis

C. Threat Analysis

D. Quantitative Analysis

9) The OWASP Foundation is an online community that produces freely-available methodologies, tools, and technologies in the field of _____ security.

A. Physical

B. Database

C. Corporate

D. Web Application

10) Which of the choices listed below is a factor to be considered when rating the severity of data integrity?

A. Modification of Data

B. Length of the Disruption

C. Single Loss Expectancy

D. Annualized Loss Expectancy

True / False Answer Key

11) An accurate assessment must transcend intangible risk severity factors as the value of the asset may extend well beyond the actual financial cost.
False

An accurate assessment must transcend tangible line-item numbers as the value of the asset may extend well beyond the actual financial cost.

12) Single Loss Expectancy (SLE) is multiplied by the Annualized Rate of Occurrence (ARO) to determine Annualized Loss Expectancy (ALE).
True

13) Most quantitative analysis approaches use a relative scale to rate risk exposures based on a set of predefined criteria established for each level.
False

Most qualitative analysis approaches use a relative scale to rate risk exposures based on a set of predefined criteria established for each level.

14) The severity rating is meant to describe the scope of the exposure, not list all the potential consequences resulting from the exposure.
True

15) Developing qualitative risk scales that address data availability are a great opportunity to design a threat assessment focus directly into the risk model.
False

Developing qualitative risk scales that address data availability are a great opportunity to design a business focus directly into the risk model.

16) Integrity severity concerns will focus primarily on unauthorized or unintended access to create, read, update, or delete data ("CRUD").
True

17) It is not uncommon to confuse factors that will affect the severity and likelihood of the risk exposure value.
False

True / False Answer Key

It is not uncommon to confuse factors that will affect the severity and likelihood of the threat / vulnerability pair.

18) An assets' Exposure Factor (EF) is multiplied by the Asset Value (AV) to determine Single Loss Expectancy (SLE).

True

19) The mission of the Common Vulnerabilities and Exposures (CVE) Program is to conduct empirical studies on critical IT infrastructure issues.

False

The mission of the Common Vulnerabilities and Exposures (CVE) Program is to identify, define, and catalog publicly disclosed cybersecurity vulnerabilities.

20) The primary purpose of data integrity severity ratings is to address varying degrees of unauthorized access to data.

True

Knowledge Assessment Answer Key

1) A(n) _____ Plan allows organizational management to reestablish leadership, allocate resources, and focus on containment and recovery.

A. Technical Continuity

B. Incident Response

C. Vulnerability Assessment

D. Emergency Operations

2) Which of the choices listed below can be considered a primary responsibility of an Emergency Operation Center's command function?

A. Gather Damage Assessments

B. Order Supplies and Services

C. Implement Allocated Resources

D. Track Recovery Personnel

3) When an Emergency Operations Center is activated, there are two primary teams: the _____ Team and the Recovery Team.

A. Investigative

B. Remediation

C. Containment

D. Logistics

4) Which of the choices listed below is a general material requirement that should be included when planning an Emergency Operations Center?

A. Funding for Overtime

B. Pre-Packaged Office Supplies

C. Human Resource Policies

D. Employee Immunizations

5) A disaster is not the time to determine what items are needed in an Emergency Operations Center; _____ will help ensure a faster recovery.

A. Hiring Policies

B. Careful Planning

C. Outsourced Solutions

D. Purchased Facilities

Knowledge Assessment Answer Key

6) Which of the choices listed below is an important factor to be considered when determining uninterruptible power supply (UPS) needs?

A. Logs to Analyze Data Events

B. Generators to Recharge the UPS

C. Battery Size and Watts Usage

D. Power Increase Plans

7) During an emergency, the _____ will communicate with teams, news media, vendors, customers, the community, and a broad range of stakeholders in the organization.

A. Command Center

B. Board of Directors

C. Senior Leadership

D. Local Managers

8) Which of the choices listed below typically falls within the scope and purpose of an Emergency Operations center?

A. Reassign Organizational Leadership

B. Initiate the Incident Investigation

C. Form the Disaster Recovery Plan

D. Minimize the Disruption of Management

9) The EOC control function involves obtaining and allocating _____ based on the needs determined by the EOC manager and leadership staff.

A. Damage Assessments

B. BCP Deviations

C. Action Plans

D. Resources

10) Which of the choices listed below represents a potential characteristic of a poorly managed disaster response scenario?

A. Large Groups Working Randomly

B. Status Board is Updated Regularly

C. Action is Directed by a Single Person

D. Everyone Knows Where to Report

True / False Answer Key

11) The mobile EOC vehicle is preloaded with everything necessary to establish an EOC, including a generator and tent for expanding the work area.

True

12) The EOC command function involves obtaining and allocating resources based on the needs determined by the EOC manager and leadership staff.

False

The EOC control function involves obtaining and allocating resources based on the needs determined by the EOC manager and leadership staff.

13) An important staffing consideration is for every member of the EOC to have a predesignated and cross-trained backup whenever possible.

True

14) Prior to sizing electrical support units such as generators and UPS, planners must know what financial requirements they will need to support in the EOC.

False

Prior to sizing electrical support units such as generators and UPS, planners must know what functions they will need to support in the EOC.

15) Most organizations convert an existing facility with specific capabilities into an Emergency Operations Center as needed.

True

16) An Emergency Operations Center must be pre-established, pre-supplied, and be an undisclosed location that is known only by senior EOC personnel.

False

An Emergency Operations Center must be pre-established, pre-supplied, and be a location that is known by everyone in advance of the time it is needed.

17) The mobile EOC vehicle is preloaded with everything necessary to establish an EOC, including a generator and tent for expanding the work area.

True

True / False Answer Key

18) If key stakeholders do not take control of the decision-making process, expensive and potentially hazardous decisions may be made in the absence of leadership.

False

If the EOC does not take control of the decision-making process, expensive and potentially hazardous decisions may be made in the absence of leadership.

19) An Emergency Operations Plan allows organizational management to reestablish leadership, allocate resources, and focus on containment and recovery.

True

20) Unless leaders can secure budgets and hire new staff during emergency situations, they are unable to effectively lead and manage the situation.

False

Unless leaders can communicate and direct their staff during emergency situations, they are unable to effectively lead and manage the situation.

Knowledge Assessment Answer Key

1) "*A business location that is used for backup in the event of an operational disaster at the normal business site and typically does not have the necessary equipment to resume prompt operations*" is the definition of a _____ site.

A. Hot

B. Reciprocal

C. Warm

D. Cold

2) Which of the choices listed below is a factor to be considered when planning team member seating needs at disaster recovery sites?

A. Suspend Location Signs from the Ceiling

B. Disperse Interactive Teams

C. Label Every Desk by Employee Name

D. Label Every Cabinet by Function

3) A _____ is critical to effective crisis management; time for recovery is decreased when teams can be assembled and tasked quickly.

A. Training Plan

B. Recovery Site

C. Conditioned Response

D. Awareness Plan

4) Which of the choices listed below is a risk to be considered when planning for the transportation of magnetic backup media?

A. Recovery Site Location

B. Temperature Change

C. Date of Data Storage

D. System to be Restored

5) The _____ Manager represents the senior leadership and must be able to manage technical and non-technical activities effectively under pressure.

A. Risk Assessment

B. Recovery Site

C. Information Technology

D. Physical Security

Knowledge Assessment Answer Key

6) Which of the choices listed below is a method of data deduplication which is typically conducted at the backup recovery site?

A. Source-Based

B. Inline

C. Target-Based

D. Post-Process

7) *"A business location that is used for backup in the event of a disruptive operational disaster at the normal business site and is typically a fully-operational commercial disaster recovery service that allows continuity of operations in a short period of time"* is the definition of a _____ site.

A. Hot

B. Reciprocal

C. Warm

D. Cold

8) Which of the choices listed below makes SMS notification systems a preferred choice of efficient mass communication?

A. Every Laptop can Receive Texts

B. High Bandwidth Usage on the Network

C. Requires the Internet to Function

D. Text Messages are Read Quickly

9) Validating _____ provides a layered testing strategy that is useful for catching errors and reducing the need to troubleshoot issues in the future.

A. Key Personnel

B. Operational Plans

C. Response Policies

D. Successful Recoveries

10) Which of the choices listed below is a standard function and purpose of an activity log at a disaster recovery site?

A. Requests for Service and Supplies

B. Status Reports to the News Media

C. An Alternative to the Gantt Chart

D. Optional Task Initiation Reporting

True / False Answer Key

11) Different levels of management and team members with different roles will need to hear the same broadcast messages with the same level of detail.

False

Different levels of management and team members with different roles will not need to hear the same broadcast messages with the same level of detail.

12) The "call tree" method for team member notification is prone to error and an inefficient option at larger scales. For this reason, it is rarely used during a crisis.

True

13) When planning for specialized issues associated with the restoration of digital communication, document alternative and practical sources of funding.

False

When planning for specialized issues associated with the restoration of digital communication, document alternative and practical communication methods.

14) Deduplication of data provides an efficient compression, single instance storage solution that reduces required storage space and speeds up recovery time.

True

15) The transport of digital backup media – although no longer common - presents specialized and high-risk challenges for the recovery site team.

False

The transport of magnetic backup media – although no longer common - presents specialized and high-risk challenges for the recovery site team.

16) Structured communication is a two-way function that is critical to reduce the chaos and confusion which exists naturally in stressful environments.

True

17) The recovery site Gantt chart is considered confidential, easy to understand, and answers the question *"who is managing the project?'*

False

True / False Answer Key

The recovery site Gantt chart is publicly posted, easy to understand, and answers the question "when will it be ready?"

18) Knowing who is on-site ensures the safety and efficiency of recovery site staff and provides documentation for post-recovery acknowledgement.

True

19) The recovery site activity log is used to record all events during the recovery and helps with pre-recovery analysis, planning, and crafting budget requests.

False

The recovery site activity log is used to record significant events during the recovery and helps with post-recovery analysis, planning, and lessons learned.

20) The Recovery Gantt Chart is used for daily status reporting and serves to identify and validate successful progress in the recovery effort.

False

The Recovery Gantt Chart is used for hour-by-hour status reporting and serves to identify and predict potential progress delays in the recovery effort.

Knowledge Assessment Answer Key

1) "*A widespread occurrence of a disease in a community that spreads quickly and affects many individuals at the same time*" is the definition of a(n) _____.

A. Epidemic

B. Influenza Outbreak

C. Pandemic

D. Common Cold

2) Which of the choices listed below is a distinctive characteristic of an epidemic that distinguishes it from a pandemic?

A. Occurs Rarely

B. Unpredictable Patterns

C. High Risk to Healthy Patients

D. Some Immunity Exists

3) The World Health Organization has divided pandemics into six phases. The _____ phase is typically the point at which a disaster is declared.

A. Animal to Human

B. Country to Country (Region)

C. Human to Human (Community)

D. Country to Country (Global)

4) Which of the choices listed below is a characteristic that is common to both influenza epidemics and pandemics?

A. A Contagious Immune System Illness

B. A Bacteria-Based Infection

C. Most Contagious in the First 1-4 Days

D. An Infection Radius up to 12 Feet

5) "*An epidemic that becomes very widespread and affects a whole region, a continent, or the world due to a susceptible population and causes a high degree of mortality*" is the definition of a(n) _____.

A. Epidemic

B. Influenza Outbreak

C. Pandemic

D. Common Cold

Knowledge Assessment Answer Key

6) Which of the choices listed below is a distinctive characteristic of a pandemic that distinguishes it from an epidemic?

A. Lower Risk to Healthy Patients

B. No Vaccines Initially Available

C. Predictable Patterns

D. Some Immunity Exists

7) When writing an epidemic / pandemic plan into the Business Continuity Plan, it is important to consider the _____ of the organizational staff.

A. Health Benefits

B. Job Titles

C. Transportation Needs

D. Family Members

8) Which of the choices listed below is an organizational policy which should be reviewed by Human Resources when creating the epidemic / pandemic plan?

A. Review the Attendance Policy

B. Identify Temporary Workers

C. Consider Mandatory Immunizations

D. Review Fixed and Permanent Shifts

9) The epidemic / pandemic _____ is based on a fluctuating business impact analysis and will vary based on unit responsibility.

A. Risk Analysis

B. Severity Rating

C. Vulnerability Assessment

D. Immunization Program

10) Which of the choices listed below is a characteristic of potentially vulnerable populations which should be considered when creating the epidemic / pandemic plan?

A. Children Over the Age of 6

B. People with Documented Allergies

C. People with Body Mass Index $\leq$ 20

D. Adults Over the Age of 65

True / False Answer Key

11) Maintain vigilance as epidemic / pandemic cases decline and utilize continuous communication until the event is officially declared over.

True

12) Common area sanitation tasks should be considered by BCP management. Request volunteers and trust those who do to manage their tasks properly.

False

Common area sanitation tasks must be assigned by BCP management; do not expect people to volunteer and hold those who are assigned accountable.

13) Provide accurate information to organizational staff well in advance of the disaster event and test SMS messages to verify current contact numbers.

True

14) Public Relations and Senior Leadership are the core of the epidemic / pandemic plan and can identify company policies impacted by risk mitigation efforts.

False

Human Resource and IT Departments are the core of the epidemic / pandemic plan and can identify company policies impacted by risk mitigation efforts.

15) Unique events require unique solutions; include organizational staff with medical and emergency services training in the epidemic / pandemic plan if possible.

True

16) When writing an epidemic / pandemic plan into the Business Continuity Plan, it is important to consider the vacation schedules of the organizational staff.

False

When writing an epidemic / pandemic plan into the Business Continuity Plan, it is important to consider the family members of the organizational staff.

17) Despite the distinction of definitions between epidemics and pandemics, an epidemic can be viewed as a localized pandemic and treated as such.

True

True / False Answer Key

18) The Centers for Disease Control and Prevention has divided pandemics into nine phases. The phase number will dictate when a disaster is declared.
False

The World Health Organization has divided pandemics into six phases. The phase number will dictate when a disaster is declared.

19) The epidemic / pandemic risk assessment is based on a fluctuating business impact analysis and will vary based on unit responsibility.
True

20) According to World Health Organization statistics, annual and seasonal epidemics worldwide typically can be expected to occur in the summer months.
False

According to World Health Organization statistics, annual and seasonal epidemics worldwide typically can be expected to occur in the winter months.

Knowledge Assessment Answer Key

1) Creating a dedicated _____ may prove to be a costly option in terms of the infrastructure, technology, personnel, and safety considerations required.

A. Risk Assessment

B. Continuity Plan

C. Data Backup

D. Recovery Site

2) Which of the choices listed below is a distinctive characteristic of the private cloud that distinguishes it from other cloud provider deployment models?

A. Formerly Termed "Intranets"

B. Resources are Leased to Customers

C. Segments are Owned by Organizations

D. Resources are Offered to the Public

3) *"A category of cloud computing services that allows customers to provision, instantiate, run, and manage a modular bundle comprising a computing platform and one or more applications"* is the definition of _____.

A. Software as a Service (SaaS)

B. Cloud Computing

C. Platform as a Service (PaaS)

D. Infrastructure as a Service (IaaS)

4) Which of the choices listed below can be considered a benefit of utilizing a cloud-based solution in business continuity planning?

A. Enables Staff to Work from One Location

B. Easily Implemented with High Reliability

C. Models are Tailored to the Providers' Needs

D. Multiple Locations Increase Redundancy

5) Even in a cloud environment, organizations cannot transfer the liability or responsibility associated with _____ of Personally Identifiable Information.

A. Storage

B. Disclosure

C. Aggregation

D. Characteristics

Knowledge Assessment Answer Key

6) Which of the choices listed below is a distinctive characteristic of the community cloud that distinguishes it from other cloud provider deployment models?

A. A Standard IT Legacy Environment

B. Resources are Owned by the Provider

C. Owners Perform Joint Tasks and Functions

D. Formerly Termed "Intranets"

7) *"A category of cloud computing services that allows customers to license software and a delivery model in which software is licensed on a subscription basis and is centrally hosted"* is the definition of _____.

A. Software as a Service (SaaS)

B. Cloud Computing

C. Platform as a Service (PaaS)

D. Infrastructure as a Service (IaaS)

8) Which of the choices listed below is a distinctive characteristic of Infrastructure as a Service (IaaS) that distinguishes it from other cloud provider service models?

A. The Provider Installs the Software

B. The Hardware is Controlled by the Customer

C. The Provider Installs the Operating Systems

D. Cloud Service is Utilized for Limited Needs

9) When discussing cloud service providers and cloud capabilities, the _____ cloud tends to be the solution to which is being referred most frequently.

A. Private

B. Community

C. Hybrid

D. Public

10) Which of the choices listed below is one of the five characteristics which are accepted as part of the NIST SP 800-145 cloud computing definition?

A. On-Demand Self-Service

B. Broad Internet Access

C. Incremental Elasticity

D. Measured Reporting

True / False Answer Key

11) Service Level Agreements contain elements of public, private, and community cloud models to varying degrees based on customer needs.
False

Hybrid clouds contain elements of public, private, and community cloud models to varying degrees based on customer needs.

12) Community clouds provide infrastructure and functionality that is owned and operated by affinity groups and similar organizations.
True

13) In the IaaS model, the cloud service provider contracts access to its infrastructure and the customer is fully responsible for its administration.
False

In the IaaS model, the cloud service provider contracts access to its infrastructure and is fully responsible for its administration.

14) Cloud services are typically offered in three models based on the capability of the provider and the needs of the customer.
True

15) Even in a cloud environment, organizations cannot transfer the liability or responsibility associated with disclosure of audit and compliance information.
False

Even in a cloud environment, organizations cannot transfer the liability or responsibility associated with disclosure of Personally Identifiable Information.

16) Managing data and its required infrastructure is neither a core function of most organizations nor is it a profit center to the organizational process.
True

17) Organizations tend to underutilize a technical resource (potential failure) or overutilize a technical resource (wasted money).
False

True / False Answer Key

Organizations tend to underutilize a technical resource (wasted money) or overutilize a technical resource (potential failure).

18) The purpose of identifying the organizations' operational state is to determine its needs, not the solutions for those needs.
True

19) Despite many functions and configurations, there are seven characteristics which are accepted as part of the NIST SP 800-145 cloud computing definition.
False

Despite many functions and configurations, there are five characteristics which are accepted as part of the NIST SP 800-145 cloud computing definition.

20) Creating a dedicated recovery site may prove to be a costly option in terms of the infrastructure, technology, personnel, and safety considerations required.
True

Knowledge Assessment Answer Key

1) Because the cloud customer and provider will each process data, they will share responsibilities and risks associated with the _____.

A. Cost

B. Location

C. Contract

D. Data

2) Which of the choices listed below is a risk associated with Personally Identifiable Information (PII) when utilizing a cloud provider platform?

A. The Provider is Financially Liable

B. The Customer cannot Recover Damages

C. The Provider Owns the Civil Liability

D. The Provider Owns the Criminal Liability

3) Although the customer is protected by the providers' acceptance of financial responsibility, legal repercussions are not the only _____ to expect.

A. Positive Benefit

B. Contract Stipulation

C. Negative Impact

D. Local Regulations

4) Which of the choices listed below represents a potential risk to customers when operating in a community cloud platform environment?

A. Resiliency Through Shared Ownership

B. No Reliability of Centralized Standards

C. Centralized Administration is not Needed

D. Shared and Distributed Costs

5) _____ can be caused when the provider goes out of business, is acquired by another interest, or ceases operation for any reason.

A. Increased Usage

B. Vendor Lock-Out

C. Contract Negotiations

D. Vendor Lock-In

Knowledge Assessment Answer Key

6) Which of the choices listed below is a factor which should be considered when researching a cloud provider to avoid the risk of vendor lockout?

A. Business Affiliations

B. Marketing Capabilities

C. Core Competencies

D. Number of Employees

7) Many potential threats posed by _____ require attenuation via the use of controls that can only be implemented by the cloud service provider.

A. Virtualization

B. External Actors

C. Natural Disasters

D. Underutilization

8) Which of the choices listed below represents a potential risk to customers when utilizing the platform of any cloud service provider?

A. Fluctuations in Share Value

B. Loss of Provider Faith and Loyalty

C. Potential Decrease in Insurance Costs

D. Damage to the Brand Image

9) Unlike a traditional legacy environment, a customer conducting operations in the cloud will not be able to conduct _____ without the provider.

A. Productivity Assessments

B. Business Processes

C. Media Campaigns

D. Local Computing

10) Which of the choices listed below is a risk factor which should be considered when conducting a cloud-specific Business Impact Analysis?

A. The Potential of Regulatory Failure

B. Decreased Ease of Data Distribution

C. The Customers' Internal Personnel

D. No New Dependencies are Created

True / False Answer Key

11) Because the cloud customer and cloud provider will each process data, they will share responsibilities and risks associated with market conditions.
False

Because the cloud customer and cloud provider will each process data, they will share responsibilities and risks associated with the data.

12) A private cloud configuration is a traditional legacy configuration of a datacenter with a variety of distributed computing capabilities.
True

13) In a hybrid cloud configuration, resources are allocated, shared, and dispersed among affinity groups.
False

In a community cloud configuration, resources are allocated, shared, and dispersed among affinity groups.

14) In a public cloud configuration, a company offers cloud services to any entity wanting to become a cloud customer and is willing to accept provider terms.
True

15) Vendor lock-in can be caused when the provider goes out of business, is acquired by another interest, or ceases operation for any reason.
False

Vendor lock-out can be caused when the provider goes out of business, is acquired by another interest, or ceases operation for any reason.

16) In the IaaS model (Infrastructure as a Service), the customer will have the most control over its resources and responsibility for asset oversight.
True

17) In the PaaS model (Platform as a Service), the customer will have all SaaS risks as well as those associated with web application responsibility and oversight.
False

True / False Answer Key

In the PaaS model (Platform as a Service), the customer will have all IaaS risks as well as those associated with platform responsibility and oversight.

18) In the SaaS model (Software as a Service), the customer will have all IaaS and PaaS risks as well as those associated with applications and access.
True

19) Many potential threats posed by virtualization require attenuation via the use of controls that can only be implemented by the cloud customers' security staff.
False

Many potential threats posed by virtualization require attenuation via the use of controls that can only be implemented by the cloud service provider.

20) The public cloud not only includes all threats associated with private clouds but also includes threats totally outside the providers' knowledge and control.
False

The public cloud not only includes all threats associated with private clouds but also includes threats totally outside the customers' knowledge and control.

Knowledge Assessment Answer Key

1) Encrypting data prior to uploading using a cryptosystem with a high work factor are components of step _____ of the Cloud Data Life Cycle (CDLC).

A. 01: Create the Data

B. 03: Use the Data

C. 04: Share the Data

D. 02: Store the Data

2) Which of the choices listed below is an option for obfuscation, masking, and anonymization that uses actual production data in its functionality?

A. Masking

B. Hashing

C. Randomization

D. Shuffling

3) With the _____ option, cloud customers using this architecture method are assigned storage space that is typically attached to a virtual machine.

A. Content Delivery Network

B. Volume Storage

C. Object-Based Storage

D. Private Cloud

4) Which of the choices listed below is a risk factor associated with Security Information and Event Management (SIEM) systems that should be considered prior to implementation?

A. Provides Trend Detection

B. Uses Dashboarding

C. All Logs are in One Location

D. Provides Automated Response

5) Adhering to Virtual Private Network, Digital Rights Management, and Data Loss Prevention requirements are components of step _____ of the Cloud Data Life Cycle (CDLC).

A. 01: Create the Data

B. 05: Archive the Data

C. 03: Use the Data

D. 02: Store the Data

Knowledge Assessment Answer Key

6) Which of the choices listed below is an option for obfuscation, masking, and anonymization that uses replacement of data with representations of data in its functionality?

A. Masking

B. Randomization

C. Hashing

D. Shuffling

7) With the _____ option, cloud customers using this architecture method are assigned storage in which their data is stored individually as opposed to files.

A. Content Delivery Network

B. Volume Storage

C. Block Storage

D. Object-Based Storage

8) Which of the choices listed below is a responsibility cloud service providers must offer cloud customers associated with the physical plant (datacenter) where the customer data is stored?

A. Use of TPM Standards for BIOS Firmware

B. Virtualization of Management Tools

C. Use of Encryption and Strong Authentication

D. Basing Security Policy on Governance

9) A _____ is a form of data caching typically located near geophysical locations of high consumer demand and data use to increase bandwidth and delivery quality.

A. Content Delivery Network

B. Volume Storage

C. Block Storage

D. Object-Based Storage

10) Which of the choices listed below is the Service Organization Control (SOC) report issued by service providers for their customers as audit assurance statements?

A. SOC 04

B. SOC 03

C. SOC 01

D. SOC 02

True / False Answer Key

11) The cloud providers' unwillingness to allow customer access to the physical plant facility applies to the customers' auditors and stakeholders as well.
True

12) An area where cloud service providers and customers may find common ground in sharing responsibilities is in the area of compliance auditing and inspections.
False

An area where cloud service providers and customers may find common ground in sharing responsibilities is in the area of security monitoring and testing.

13) In all cloud service models, the customer and their users will need to access and modify the data at various levels using various permissions and controls.
True

14) The Intrusion Detection System is a large attack surface and offers many potential attack vectors to malicious actors if not secured correctly.
False

The operating system is a large attack surface and offers many potential attack vectors to malicious actors if not secured correctly.

15) The cloud provider must apply the proper security controls according to the customers' relevant regulatory industry frameworks and planned usage.
True

16) In addition to securing the hardware components, the cloud service provider must ensure that contractual documents are equally protected for the customer.
False

In addition to securing the hardware components, the cloud service provider must ensure that the logical elements are equally protected for the customer.

17) The physical plant of the datacenter will include the facility campus, all physical components, and the services and personnel that support them.
True

True / False Answer Key

18) Some degree of adversarial relationship exists between the customer and the cloud services provider because they reside in different physical locations.
False

Some degree of adversarial relationship exists between the customer and the cloud services provider because they have differing business goals.

19) An egress monitoring solution will examine data leaving the production environment and react based on preestablished rules and parameters.
True

20) The methods and locations chosen to effectively manage encrypted keys impacts the overall value and benefit of data in several ways.
False

The methods and locations chosen to effectively manage encrypted keys impacts the overall risk to data in several ways.

Knowledge Assessment Answer Key

1) A great deal of the difficulty in managing the legal aspects of cloud computing stems from the design and distribution of the _____ themselves.

A. Cloud Assets

B. Customer Locations

C. Logistic Centers

D. Governing Agencies

2) Which of the choices listed below is a characteristic of the ISO 31000:2009 risk management framework that distinguishes it from other related risk frameworks?

A. Supersedes the Older "C&A" Model

B. Identifies 35 Types of Risks

C. Not Accepted in International Markets

D. Explicitly Address Uncertainty Factors

3) Policies are a foundational element of governance and risk management programs and ensure companies operate within their chosen _____.

A. Industries

B. Risk Profiles

C. Market Area

D. Geographic Location

4) Which of the choices listed below is a characteristic of the NIST SP 800-39 risk management framework that distinguishes it from other related risk frameworks?

A. The Program is Transparent and Inclusive

B. Considers the Human and Cultural Factors

C. Relies Heavily on Automation

D. 8 Security Risks are Based on Likelihood

5) The variety and vagaries of multijurisdictional _____ make the regulatory entities, stakeholders, and their combined input complicated for cloud services.

A. Policies

B. Customs

C. Laws

D. Clients

Knowledge Assessment Answer Key

6) Which of the choices listed below is a characteristic of the ENISA risk management framework that distinguishes it from other related risk frameworks?

A. Not Accepted in International Markets

B. Identifies 35 Types of Risks

C. Must be Dynamic, Iterative, and Responsive

D. Relies Heavily on Automation

7) Once an organizational policy has been _____, it must be published and disseminated to those individuals affected by the policy.

A. Edited for Errors

B. Peer Reviewed

C. Initially Drafted

D. Formally Accepted

8) Which of the choices listed below is a characteristic of a Service Level Agreement that distinguishes it from the cloud computing contract in which it resides?

A. Defines Numerical Metrics

B. Stipulates Penalties

C. Describes Responsibilities

D. Outlines Services Offered

9) It is vitally important that both the customer and the cloud provider focus on relevant _____ issues and the challenges of cloud computing.

A. Risk Management

B. Service Agreement

C. Product Marketing

D. Business Development

10) Which of the choices listed below is a factor to be considered while prioritizing jurisdictions when developing organizational cloud computing policies?

A. Identify Jurisdictions with Minimal Impact

B. Crafting Individual Policies is an Option

C. The Smallest Residence of End Clientele

D. The Most Bearing on Organizational Policy

True / False Answer Key

11) It is important that the cloud provider consider all possible situations and risks associated with cloud business processes, requirements, and potential outcomes.

False

It is important that the cloud customer consider all possible situations and risks associated with cloud business processes, requirements, and potential outcomes.

12) While both the contract and SLA may contain numerical values, the SLA will expressly include metrics to determine if contractual goals are being met.

True

13) To understand whether business processes and procedures are effective, it is important to identify metrics that accurately reflect the RM program goals.

False

To understand whether control mechanisms and policies are effective, it is important to identify metrics that accurately reflect the RM program goals.

14) The *European Union Agency for Network and Information Security* (ENISA) is widely accepted within Europe, but not globally accepted like ISO standards.

True

15) The NIST SP 800-39 risk management framework is a methodology for handling all risk in a holistic, comprehensive, and continual manner.

True

16) The ISO 31000:2009 is an international standard focused on designing, implementing, and reviewing vulnerability assessment practices and processes.

False

The ISO 31000:2009 is an international standard focused on designing, implementing, and reviewing risk management practices and processes.

17) When discussing policy matters with stakeholders, consider that most will not have a complete grasp of cloud computing technology and its implications.

True

True / False Answer Key

18) Once an organizational policy has been initially conceived and drafted, it must be published and disseminated to those individuals affected by the policy.
False

Once an organizational policy has been formally accepted, it must be published and disseminated to those individuals affected by the policy.

19) Identifying and engaging relevant stakeholders is vital to the success of any cloud computing discussions, programs, or planned projects.
True

20) Frameworks are a foundational element of governance and risk management programs and ensure companies operate within their chosen risk profiles.
False

Policies are a foundational element of governance and risk management programs and ensure companies operate within their chosen risk profiles.

Knowledge Assessment Answer Key

1) The _____, also known as the *Financial Services Modernization Act* (FSMA) of 1999, was created to allow banks and financial institutions to merge.

A. Health Insurance Portability & Accountability Act

B. Sarbanes-Oxley Act

C. Stored Communication Act

D. Gramm-Leach-Bliley Act

2) Which of the choices listed below is a distinctive characteristic of administrative law that distinguishes it from criminal, civil, and international law?

A. Created by Executive Decision and Function

B. Enacted by State Legislatures

C. Strictly for Private Entities

D. Addresses Trade Regulations

3) The _____ was enacted in 2002 as an attempt to prevent unexpected financial collapse due to fraudulent accounting practices and poor oversight.

A. Health Insurance Portability & Accountability Act

B. Stored Communication Act

C. Sarbanes-Oxley Act

D. Gramm-Leach-Bliley Act

4) Which of the choices listed below is a distinctive characteristic of criminal law that distinguishes it from administrative, civil, and international law?

A. Created by Executive Decision and Function

B. Created by Federal and State Laws

C. Addresses Breach of Contract

D. Addresses Tariff Structures

5) _____ involves all legal matters where the government conflicts with any person, group or entity that violates various laws and/or statutes.

A. Civil Law

B. Criminal Law

C. Administrative Law

D. International

Knowledge Assessment Answer Key

6) Which of the choices listed below is a distinctive characteristic of civil law that distinguishes it from administrative, criminal, and international law?

A. Created by Executive Decision and Function

B. Federal Courts Typically Handle the Cases

C. Addresses Tort Laws

D. Addresses Intellectual Property Law

7) The Doctrine of _____ is a term used to describe the processes associated with determining what legal jurisdiction will hear disputes.

A. Proper Law

B. Res Juriscata

C. Transferred Intent

D. Judicial Discretion

8) Which of the choices listed below is a distinctive characteristic of international law that distinguishes it from administrative, criminal, and civil law?

A. Created by Executive Decision and Function

B. Enacted by State Legislatures

C. Addresses Patents and Trademarks

D. Practiced Customs are Accepted as Law

9) The _____ supersedes the *Data Directive* and ends the *Safe Harbor* program, replacing it with a program called *Privacy Shield*.

A. Transparency Regulation

B. Technology Regulation

C. Data Regulation

D. Privacy Regulation

10) Which of the choices listed below is a reason for the *Doctrine of Proper Law* to be considered by legal entities when preparing to hear cases in a court of law?

A. Determine the Legal Jurisdiction

B. Determine the Violated Laws

C. Determine the Affected Parties

D. Determine Fines and Penalties

True / False Answer Key

11) Decentralized data and its movement, storage, and processing across geographic boundaries creates complex challenges for forensic investigations and standards.
True

12) All contracts need to be secured, tracked, and monitored from the time it is acquired and identified for investigative purposes until the case is closed.
False

All evidence needs to be secured, tracked, and monitored from the time it is acquired and identified for investigative purposes until the case is closed.

13) *Electronic Discovery* (eDiscovery) refers to the process of identifying and obtaining evidence for criminal prosecutorial or civil litigation purposes.
True

14) Due to the centralized nature of cloud computing, many geographic similarities legal parity may present critical personal and data privacy issues.
False

Due to the decentralized nature of cloud computing, many geographic disparities and lack of parity may present critical personal and data privacy issues.

15) The *Privacy Regulation* supersedes the *Data Directive* and ends the *Safe Harbor* program, replacing it with a program called *Privacy Shield*.
True

16) The United States' *Data Protection Directive* of 1995, referred to as the "Data Directive," was the first major national data privacy law.
False

The European Union's *Data Protection Directive* of 1995, referred to as the "Data Directive," was the first major European Union data privacy law.

17) The *Doctrine of the Proper Law* is a term used to describe the processes associated with determining what legal jurisdiction will hear disputes.
True

True / False Answer Key

18) International law is a body of law that affects most people; it is not created by legislatures, but by executive decision and function.
False

Administrative law is a body of law that affects most people; it is not created by legislatures, but by executive decision and function.

19) Criminal law involves all legal matters where the government conflicts with any person, group or entity that violates various laws and/or statutes.
True

20) Administrative law determines how to settle disputes and manage relationships between sovereign, recognized countries and their respective entities.
False

International law determines how to settle disputes and manage relationships between sovereign, recognized countries and their respective entities.

Knowledge Assessment Answer Key

1) The best way to avoid arguments in a business relationship is to _____ the parties' expectations ahead of time, then understand and agree to them.

A. Write Down

B. Negotiate

C. Reject

D. Demand

2) Which of the choices listed below is a distinctive characteristic of Prime Clauses that distinguishes them from other sections in general legal contract structures?

A. Includes Service Level Agreements

B. Includes Introductions and Recitals

C. Addresses Nondisclosure and Confidentiality

D. Used for Defining the Professional Service

3) Cybersecurity and Information Technology contracts can be organized into three groups: *prime clauses*, _____, and *boilerplate clauses*.

A. Statements of Work

B. General Clauses

C. Service Level Agreements

D. Amendments

4) Which of the choices listed below is a component of an enforceable contract which must be present for the document to be considered legal and acceptable?

A. Filing Time Stamp

B. Letterhead

C. Mutuality of Obligation

D. Notary Seal

5) You cannot predict what issues may arise in a business relationship, so you cannot know when the content of _____ clauses will become vital.

A. Severability

B. General Clauses

C. Boilerplate

D. Prime

Knowledge Assessment Answer Key

6) Which of the choices listed below is a distinctive characteristic of General Clauses that distinguishes them from other sections in general legal contract structures?

A. Includes Statements of Work

B. Includes Service Level Agreements

C. Identifies Choice of Law and Courts

D. Includes Amendments

7) In negotiation, include a _____ in your contract — a guarantee that one side can match any offer that the other side later receives.

A. Post-Settlement

B. Contingent Agreement

C. Disclaimer

D. Matching Right

8) Which of the choices listed below is a reason to seek post-settlement settlements when negotiating the details of legal contracts and formal agreements?

A. Can Lead to Increased Mutual Trust

B. Considers Alternative Dispute Resolution

C. Talks can Continue Despite Disagreement

D. Addresses "Right of First Refusal"

9) When reviewing contracts and service level agreements, consider the use of _____ to help ensure a consistent approach for evaluation.

A. Standardized Checklists

B. Peer Review

C. Second Opinions

D. Legal Dictionaries

10) Which of the choices below is a distinctive characteristic of Boilerplate Clauses that distinguishes them from other sections in general legal contract structures?

A. Includes Change Order Procedures

B. Includes Schedules and Milestones

C. Addresses Limitation of Liability

D. Includes Introductions and Recitals

True / False Answer Key

11) Customers should protect themselves against unfair contracts and ensure all promises and commitments address the expectation of service needed.
False

Customers should protect themselves against unclear descriptions and ensure all promises and commitments address the expectation of service needed.

12) The needs of the organization do not always align with the desires of the service provider. For that reason, measurements should be chosen by the client.
True

13) The true costs associated with changing service providers may not always be immediately clear. For that reason, advance payment is a requirement.
False

The true costs associated with changing service providers may not always be immediately clear. For that reason, written clarification is a requirement.

14) Words are important, and clear definitions even more so. Ensure what is being stated in the contract accurately reflects the reality of the situation.
True

15) Most disagreements arise from contracts that are not clear, complete, and that do not express the agreement as it was understood by the courts involved.
False

Most lawsuits arise from contracts that are not clear, complete, and that do not express the agreement as it was understood by the parties involved.

16) The *Boilerplate Clauses* include a set of terms usually placed at the end of a contract and are also used for introductory material as well.
True

17) The one characteristic shared by all elements of *General Clauses* is that they generate the most agreement, consensus, and generous terms.
False

True / False Answer Key

The one characteristic shared by all elements of *General Clauses* is that they generate the most disagreement, debate, and compromise.

18) When formulating the initial contract, you can — and should — specify what will happen if either side violates the terms of the contract.
True

19) In negotiation, parties often reach an impasse because they have similar beliefs about the likelihood of future events being discussed at the time.
False

In negotiation, parties often reach an impasse because they have different beliefs about the likelihood of future events being discussed at the time.

20) Each party has a choice about whether to enter and agree to a contract; neither side owes the other any special considerations or terms during the process.
True

Knowledge Assessment Answer Key

1) As with all things in the Business Continuity program, begin setting goals for the testing plan by referring to the _____.

A. Business Impact Analysis

B. Risk Management Framework

C. Corporate Security Policy

D. Local Emergency Management

2) Which of the choices listed below is a distinctive characteristic of *Checklist Testing* that distinguishes it from other progressions of BCP / DRP testing scenarios?

A. Tests the Logical Grouping of Plans

B. Utilized to Determine a Realistic RTO

C. Incorporates Mid-Exercise Problem Injections

D. The First Level of Plan Error Checking

3) Testing can be structured to focus on a plans' strengths and gloss over weaknesses. Third-party _____ seeks to overcome this issue.

A. Involvement

B. Certification

C. Outsourcing

D. Opinions

4) Which of the choices listed below is an example of a planned external event which can be used to facilitate improvisational BCP / DRP testing opportunities?

A. Natural Disasters

B. Civil Unrest

C. Local Government Disaster Drills

D. Organizational Emergencies

5) Use blogs, newsletters, posters, and email tips to maximize the messaging of training. _____ training tends to be much more engaging.

A. Well-Attended

B. Mandatory

C. Highly-Interactive

D. Quarterly

Knowledge Assessment Answer Key

6) Which of the choices listed below is a distinctive characteristic of *Parallel Testing* that distinguishes it from other progressions of BCP / DRP testing scenarios?

A. Used to Create 100% Confidence in the Plan

B. Includes Relevant Personnel and Equipment

C. Ensures the Recovery Plan is Understandable

D. The First Level of Plan Error Checking

7) Realistic testing scenarios based on _____ tend to generate more "buy in" from BCP testing participants and encourage active engagement.

A. Leadership Suggestions

B. Popular Shows

C. Employee Experience

D. Recent Events

8) Which of the choices listed below is an example of a BCP / DRP team member who would be expected to attend any organizational BCP testing exercise?

A. Business Continuity Manager

B. Court Recorder

C. Corporate Sponsors

D. Unsolicited Volunteers

9) Whenever an incident occurs that is covered by the BCP, conduct a(n) _____ on the next workday after the recovery with participants involved.

A. After-Action Review

B. Press Conference

C. Management Presentation

D. Training Seminar

10) Which of the choices listed below is a distinctive characteristic of *Fail-Over Testing* that distinguishes it from other progressions of BCP / DRP testing scenarios?

A. The First Level of Plan Error Checking

B. May Utilize Recovery Site Systems

C. Ensures the Recovery Plan is Understandable

D. A Full Shutdown of the Primary Location

True / False Answer Key

11) To maximize benefits to the organization while minimizing costs, develop a written testing strategy for the BCP that can be planned and revised in advance.
True

12) The best BCP testing results come from a clear explanation of the responsibilities of unit managers and providing training to demonstrate the plan expectations.
False

The best BCP testing results come from a clear explanation of the responsibilities of team members and providing training to demonstrate the plan expectations.

13) The scenarios selected should be focused on the types of problems that a specific group of people within the organization are likely to encounter.
True

14) The 6 progressions of BCP testing allow for exponential improvements over time and a smaller variety of testing to be conducted less often.
False

The 5 progressions of BCP testing allow for incremental improvements over time and a wider variety of testing to be conducted more often.

15) When testing exercise outcomes fall short of expectations, do not blame participants. Blame and fault are not productive and discourage participation.
True

16) Unplanned internal events can create unique opportunities to execute testing scenarios that are relevant to various aspects of the business operation.
False

Planned external events can create unique opportunities to execute testing scenarios that are relevant to various aspects of the business operation.

17) The objective of a *Fail-Over Test* is to create a full interruption of organizational systems to definitively test the BCP implementation and the personnel involved.
True

True / False Answer Key

18) The objective of a *Table-Top Test* is to create and work through a BCP that is standardized and understood by all participants involved in the process.
False

The objective of a *Table-Top Test* is to train specific team members, identify omissions, and raise general awareness of the BCP and its overall goals.

19) A *Structured Walk-Through* is conducted in the actual test environment and typically focuses on specific IT systems, personnel, and potential challenges.
True

20) The objective of a *Parallel Test* is to create and work through a BCP that is standardized and understood by all participants involved in the process.
False

The objective of a *Parallel Test* is to evaluate the BCP with actual systems and relevant personnel without the risk of interrupting the organizations' operation.

Acronyms and Abbreviations

ALE	Annualized Loss Expectancy
API	Application Programming Interface
ARO	Annual Rate of Occurrence
AV	Antivirus / Asset Value
BCP	Business Continuity Plan
BIA	Business Impact Analysis
BIOS	Basic Input and Output System
BYOD	Bring Your Own Device
C2	Command and Control
CASB	Cloud Access Security Broker
CDC	Centers for Disease Control and Prevention
CDN	Content Delivery Network
CFAA	Computer Fraud and Abuse Act of 1986
CIAA	Confidentiality, Integrity, Availability, and Accountability
CLDC	Cloud Data Life Cycle
CRM	Customer Relationship Management
CRUD	Create, Read, Update, or Delete
CVE	Common Vulnerabilities and Exposures
DevOps	Software Development and IT Operations
DHHS	Department of Health and Human Services
DLP	Data Loss Prevention
DRM	Digital Rights Management
DRP	Disaster Recovery Plan
ECPA	Electronic Communications Privacy Act of 1986
EF	Exposure Factor
EMS	Emergency Management Services
ENISA	European Union Agency for Network and Information Security
EOC	Emergency Operations Center
EU	European Union
FAIR	Factor Analysis of Information Risk
FIPS	Federal Information Processing Standard
FRAAP	Facilitated Risk Analysis and Assessment Process
FSMA	Financial Services Modernization Act of 1999 (same as GLBA)
GDPR	General Data Privacy Regulation of 2016
GLBA	Gramm-Leach-Bliley Act of 1999
HIPAA	Health Insurance Portability and Accountability Act of 1996

HVAC	Heating, Ventilation, and Air Conditioning
IaaS	Infrastructure as a Service
IAM	Identity and Access Management
IDS	Intrusion Detection System
IPS	Intrusion Prevention System
IPsec	Internet Protocol Security
ISO	International Organization for Standardization
IT	Information Technology
LEF	Loss Event Frequency
MAD	Maximum Allowable Downtime
MESO	Multiple Equivalent Simultaneous Offers
NIST SP	National Institute of Standards and Technology Special Publication
O/S	Operating System
OCR	Office of Civil Rights
OCTAVE	Operationally Critical Threat, Asset, and Vulnerability Evaluation
OWASP	Open Web Application Security Project®
PaaS	Platform as a Service
PHI	Personal Healthcare Information
PII	Personally Identifiable Information
PLM	Probable Loss Magnitude
RMF	Risk Management Framework
RPO	Recovery Point Objective
RTO	Recovery Time Objective
SaaS	Software as a Service
SCA	Stored Communication Act of 1986
SEC	Securities and Exchange Commission
SIEM	Security Information and Event Management
SLA	Service Level Agreement
SLDC	System Development Life Cycle / Software Development Life Cycle
SLE	Single Loss Expectancy
SMS	Short Message Service
SOC	System and Organization Controls
SOX	Sarbanes-Oxley Act of 2002
TPM	Trusted Platform Module
UPS	Uninterruptable Power Supply
VM	Virtual Machine
VPN	Virtual Private Network
WHO	World Health Organization

Made in the USA
Columbia, SC
22 November 2021

49254168R00159